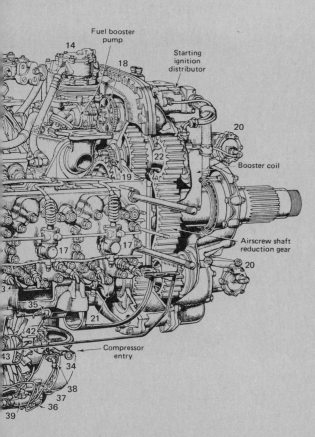

Variable Gear

1. Disk-loading spring.
2. Swinging shaft.
3. Oil-pressure feed gallery to centershaft disks.
4. Gearbox breather, with matrix separating oil from air.
5. Gear ratio indicator.

Wheelcase

6. Hand-turning gear.
7. Epicyclic reduction gear.
8. Turbine governor and tachometer drives.
9. Viscous crankshaft damper.

Top Cover

10. Fuel-rack cross shaft.
11. Turbine over-speed governor.
12. Turbine tachometer generator.
13. Crankshaft tachometer generator.

Fuel System

14. De-aerator.
15. Filters (4).
16. Injection pumps (2).
17. Injectors (12).

Front Casing

18. Booster-pump drive.
19. Injection-pump drive take-off.
20. Torquemeters.
21. Flexible-shaft front coupling.
22. Layshaft reduction gear.

Cylinder Block

23. Igniter plugs.
24. Dry liner.
25. Induction ports.
26. Exhaust ports.

27. Induction-port deflectors.
28. Coolant distribution gallery.
29. Coolant return to outlet.
30. Coolant outlet.
31. Hot-crown piston.
32. Induction trunk.
33. Coolant pump, delivery to block.

Compressor

34. Entry annulus, with struts and bullet de-iced by 12th-stage air.
35. Air pitot pressure.
36. Air static pressure.
37. Air temperature.
38. Oil to front bearing.
39. Sump.
40. Oil return to scavenge pump.
41. 12th-stage supply to de-ice intake.
42. Variable-incidence intake vanes.
43. Filter of 4th-stage air to turbine and compressor bearings.
44. Air to cool compressor bearings.
45. Oil to rear bearing.
46. Bifurcated delivery trunk.
47. Coupling shaft to turbine.
48. Compressor-oil scavenge pump.

Turbine

49. Exhaust-pipe pressure.
50. 4th-stage compressor air to front and rear bearing.
51. 12th-stage air to thrust-balance piston.
52. Thrust-balance piston.
53. 12th-stage disk-cooling air.
54. Labyrinth seal pressurized by 12th-stage air.
55. Circulation space for 4th-stage bearing-cooling air.
56. Tail-pipe pressure.

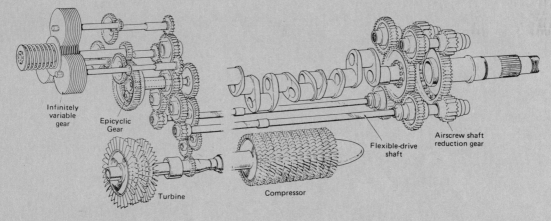

AIRCRAFT PISTON ENGINES

AIRCRAFT PISTON ENGINES

From the Manly Baltzer to the Continental Tiara

HERSCHEL SMITH

McGraw-Hill Book Company

New York   St. Louis   San Francisco   Auckland
Bogotá   Hamburg   Johannesburg   London   Madrid
Mexico   Montreal   New Delhi   Panama   Paris
São Paulo   Singapore   Sydney   Tokyo   Toronto

McGRAW-HILL SERIES IN AVIATION
DAVID B. THURSTON: Consulting Editor

RAMSEY • *Budget Flying* (1981)
SACCHI • *Ocean Flying* (1979)
SMITH • *Aircraft Piston Engines* (1981)
THOMAS • *Personal Aircraft Maintenance* (1981)
THURSTON • *Design for Flying* (1978)
THURSTON • *Design for Safety* (1980)
THURSTON • *Home-Built Aircraft* (1981)

Library of Congress Cataloging in Publication Data
Smith, Herschel H
   Aircraft piston engines.

   (McGraw-Hill series in aviation)
   Bibliography: p.
   Includes index.
   1. Airplanes—Motors. I. Title. II. Series.
TL701.S5   629.134′352   80-22606
ISBN 0-07-058472-9

1234567890   VHVH   8987654321

The editors for this book were Jeremy Robinson and Geraldine Fahey, the designer was Mark E. Safran, and the production supervisor was Thomas G. Kowalczyk. It was set in Souvenir Light by J.M. Post Graphics.

Printed and bound by Von Hoffmann Press, Inc.

To Ruth Smith, who passed with high honors the very severe test of a wife's patience that is set by having one's husband write a book.

Contents.

Preface.

I wrote this book because I got tired of waiting for someone else to do it.

There are many very good aviation historical books in existence, but they seldom have much to say about the powerplants that make the whole thing possible. Over the years I have read most of these books and have, increasingly, hungered for more information about the engines involved. I would wonder: what is an R-2000? What were the differences among all the different Gipsies? How much alike were the Spitfire's Merlin and the Messerschmitt's Daimler-Benz? What was it about the Wright Whirlwind that changed the Atlantic crossing from a suicide mission to a feasible operation? Why did France, Britain, and Germany use such widely differing types of engines in the First World War? Why did the Diesel aircraft engine ultimately fail? What has caused the present worldwide domination of the horizontally opposed type of engine?

Hopefully this book will answer most such questions. There are some valuable engine histories in existence, but they are hard to come by, they are incomplete, and they are not always as accurate as one would like them to be. The trouble may be that most of them are written by experts. I am no engine expert; I am simply a technology fan in the same sense that many people are sports fans. But perhaps this is for the best, because, if I knew more, I might have relied too much on my own recollection and done less cross-checking. As it is, I began knowing just enough to tackle the job; most of the information here is the result of research.

Careful research sometimes yields conflicting data. The reader should go slow in assuming that a figure that is different from what has been seen elsewhere is necessarily wrong. The horsepower figures for many engines vary greatly over time, with altitude rating, with different "dash numbers," and for different applications. So do dates and weights. Many of the major engines deserve monographs of their own for it is only in a monograph that all the permutations of a given design can be explored. To get broad coverage, as I have tried to do, some depth must be sacrificed.

In some cases it has been easy to determine that certain engines have been outstandingly good or bad. In others, I have had to decide between conflicting opinions. For some engines, good data have been available; for others, data have been minimal or doubtful. In the end, a historian must make judgments.

Sources and acknowledgments.

Probably the most important sources of data have been *Jane's All the World's Aircraft,* over the years, and Glenn D. Angle's two remarkable works—the 1921 *Aircraft Engine Cyclopedia* and the 1939 *Aerosphere.* Engine histories by Heron and Schliaffer, Taylor, and Setright have been helpful. Various aircraft historical works have helped in cross-checking and making deductions. The main source for reducing the inscrutability of Japanese engines has been Rene J. Francillon's *Japanese Aircraft of the Pacific War.* The Smithsonian *Annals of Flight* monographs—*Aircraft Propulsion* (Taylor), *Langley's Aero Engine of 1903* (Meyer), *The Curtiss D-12 Engine* (Byttebeir), *The Liberty Engine, 1918–1942* (Dickey), and *The First Airplane Diesel Engine: Packard Model DR-980 of 1928* (Meyer)—have been invaluable. *By Jupiter,* the Fedden biography, was an eye-opener. And, finally, publications like *Air Progress, Sport Aviation, Flight,* and the old *Aero Digest* have been helpful. Rolls-Royce, Allison, Lycoming, and Continental have contributed material.

Equally important has been the help from people, some of whom could undoubtedly have done a better job than I have had they been free to take it on. Without the help of Robert B. Meyer, Jr., Curator of Aircraft Propulsion at NASM-Smithsonian, and Harvey H. Lippincott, Corporate Archivist at Pratt and Whitney, the difficulties would have been much greater. The novelist and aero historian Len Deighton deserves special thanks for pushing the project past top dead center and getting it turning over. Mr. Merrill Stickler, Curator of the Glenn Curtiss Museum of Local History, and his staff were especially patient and helpful, notably with regard to the career of Charles B. Kirkham. Mike Evans of Rolls-Royce was very helpful. Dr. Peter Williamson, thanks largely to Alden Sherman's vouching for me, helped enormously by lending me his valuable collection of *Jane's* books. Everett Cassagneres, the Ryan Aircraft historian, contributed Wright and Kimball information. Professor Emeritus August Rogowski of MIT gave his time generously to strengthen my understanding of basic engine relationships. Mr. E. Paul Jones, Head Librarian of the Business and Technology Room at Bridgeport's Burroughs Library, was most helpful and patient. And, overall, McGraw-Hill's Jeremy Robinson and Consulting Editor David Thurston are responsible for whatever degree of readability and order the book possesses.

ONE

An introduction to the aircraft piston engine.

Almost from the beginning of powered flight, reliability has been the distinguishing characteristic of a good airplane engine. One must say "almost" because the first powered flights were only from one end of a large field to the other, and no great harm resulted if the engine stopped. All that the first fliers asked—and, in the years before 1908, it was quite a lot to ask—was for an engine light enough in relation to its power to get their machines into the air. But as soon as they began to venture across country, a willingness to keep running until shut off became the most desirable engine attribute.

Reliability

In the pursuit of reliability, as in most engineering developments, there are two roads to success. One is by the refinement of existing designs and concepts; the other is by the introduction of new and superior ideas. Each approach has its advantages. On the one hand, it is an engineering adage that fully developed examples of well-tried concepts invariably work better than the first products of new thinking, but, on the other, there is a limit to what can be done with any design concept, and new directions must be explored if progress is to be made. For example, years of hard work had brought the best water-cooled engines to a plateau of dependability by the mid-twenties. Further efforts devoted to improving water-cooled engines were bound to yield diminishing returns. In contrast to this, adoption of air cooling promised an immediate 25 percent reduction in forced landings, this being the proportion of engine failures known to be traceable to plumbing problems.

When the air-cooled radial engine emerged, it more than fulfilled this promise. While the radial could not have achieved the success that it did had it not incorporated much of the advanced technology of the latest water-cooled engines, the major advance came from the new and different concepts that it embodied [Fig. 1-1(a) and (b)].

FIG. 1-1(a)   Curtiss D-12, 400 hp—America's best water-cooled engine, 1921 to 1928.

On the piston engine so many things must work well at all times that a great deal of development has been required to bring it to a state of high reliability. The earliest aircraft engines were appallingly prone to mechanical failures. Since the technology of the time was incapable of getting adequate power from a relatively small engine, the only way to get a usable power-to-weight ratio was to build large engines with every part made as light as possible. Unfortunately, the knowledge of just how light things could be made while still having adequate strength was also lacking. The result was that any part that could break, burn out, or fall off was likely to do so. A great deal had to be learned before engines light enough for aircraft became mechanically sound.

Nevertheless there is a certain technical straightforwardness to developing an engine as a well-integrated collection of properly made mechanical parts. It was the engines' accessories that provided the more difficult challenge. Consider the ignition system. The spark that fired the fuel mixture came from a magneto, an invention of the Devil if there ever was one. It gives a weak spark at low speeds, and, considering the state of knowledge about insulating materials before 1920, it is hardly surprising that occasionally it gave no spark at all.

We in the United States had so little confidence in our ability to build battleworthy magnetos in 1917 that we equipped our Liberty engines with automobile-type coil-and-distributor ignition systems. As early as 1913, aircraft engines were being built with *twin ignition*—a pair of duplicated systems— and this remains the standard practice. While it is a reasonable assumption that one system will get you home if the other one quits, clearly the use of redundant systems is a tacit admission that a reliability problem still exists.

The other critical ignition component was the spark plug. In the early days there was a bewildering variety of designs, many of which would work well enough on one engine but not on another. Then, just when spark plug design had settled down somewhat, around 1921, radio came along to start the problem all over again: plugs had to be developed that didn't cause radio interference. Even today there is occasional plug trouble, and changing plugs is probably the most common minor repair for lightplane engines.

Another persistently troublesome accessory was the carburetor, the device that mixes air with gasoline to create an explosive mixture. The design of carburetors has always been a blend of art, science, and witchcraft. The basis of normal carburetors is a *venturi,* that is, a necked-down section in an air pipe. Pass air through a venturi and a partial vacuum is created; place a fuel jet in a venturi and the vacuum will suck fuel into the airstream. The more air that passes the venturi, the stronger the vacuum that is created and the more fuel that is sucked in. So there is a tendency for the air-to-fuel proportions to remain constant over a wide range of throttle openings. This is all very well, but the evaporating fuel makes the venturi carburetor an effective refrigerating device, and this is the main source of carburetor unreliability. On cool, damp days the venturi can actually clog up with ice. When this happens, the engine becomes very quiet.

The nominal cure is to preheat the incoming air by passing it over the hot exhaust pipe. This has two faults. The lesser is that using heat this way reduces the engine's power because heated air is less dense than cool air and picks up less fuel as it passes through the venturi. The more serious fault was stated in one manufacturer's operating manual

FIG. 1-1(*b*) Wright J-5 Whirlwind, 220 hp, 1925. After J-5s powered the first nonstop Atlantic and Pacific crossings, the water-cooled engine went into a permanent decline in the United States. (*Courtesy of NASM-Smithsonian.*)

as "When you need heat there isn't any." This means that if the application of heat is left too long, the engine will have slowed down so much that the exhaust pipe will be too cool to do any good. The pilot or flight engineer must monitor the engines closely when the temperature is between 40 and 50°F (5 and 11°C) and must apply heat at the first indication of power loss.

The eventual solution to carburetor icing was to eliminate the venturi carburetor. This was done in two stages. The first was the *injection,* or *pressure, carburetor* of around 1936. This device sprayed fuel under pressure into the airstream, an eminently sensible approach that, however, called for some sophisticated arrangements for holding the fuel-and-air mixture to its proper proportions. The final stage was to inject the fuel through small nozzles at each cylinder. Even here, we are somewhat short of perfection; a fuel-injected engine is sometimes harder to start than a carbureted one.

Another source of unreliability, as mentioned earlier, was the plumbing associated with water-cooling systems. We in the United States gave up water cooling almost completely when air-cooled engines of adequate power became available. The French and the Italians solved the reliability problems of water-cooled engines to a degree that permitted some extraordinary over-water long-distance flights to be made, as, to some degree, did the Russians. The Germans, too, preferred to plug away at solving the problems of liquid cooling; they never designed a good large air-cooled radial, and the ones they did build were based on American designs. The British, on the other hand, were successful both in achieving good reliability with liquid-cooled engines and in building good air-cooled types.

Power

Closely following on improvements in reliability came increases in output. Aircraft other than trainers justify their existence by their ability to carry loads and to go fast and far, capabilities that are enhanced in direct proportion to the amount of power available. There are two ways to build a more powerful engine. One is to extract more power from an engine of a given size, effective engine size being the volume displaced by the pistons in their travel. The other is to make a larger engine of the same power per cubic inch (or per liter) of displacement. Increasing the *specific output,* as the power per unit of displacement is called, requires advances in the state of the art, both in obtaining the extra power and in getting an engine to stand up to the higher stress and heat that goes with it. Designing a larger engine of the same specific output is often simpler, involving little more than scaling up an existing design or making an engine with more cylinders of a design that has worked well in the past. Hence, upward steps in power have tended to come first in the form of engines that have been larger but not especially advanced in design over previous ones. More advanced designs come along later to offer the same power increases in smaller sizes and, usually, lighter weight.

Engines can be enlarged by using larger cylinders or by using more cylinders of the same size. More cylinders mean more complications, but larger ones sometimes cause problems. This has been especially true with respect to air-cooled engines. When a cylinder design is scaled up, the power and heat generated all increase by the cube of the enlargement factor, but the surface area and the ability to dissipate heat increase only by the square of the enlargement factor. Thus, cooling arrangements that were barely adequate for the original design are apt to be inadequate for the enlarged version.

A less obvious problem with scaled-up engines is that vibration increases disproportionately. The British ran into this when, late in the First World War, they enlarged one of their sixes to the same displacement as existing eights and twelves. Apparently the idea was that the enlarged engine would go into a stronger airframe that would be able to cope with the worsened vibration, but the shake factor went up all out of proportion to the increase in size and power. Curtiss-Wright was luckier when, in the mid-thirties, it came out with the first engine to produce 1000 hp from only 9 cylinders. Just in time, a new and superior type of engine-mounting device, called the "Dynafocal," became available. The Dynafocal mount allowed considerable freedom of movement in one direction while still providing the necessary restraint in others.

Its use brought the perceived vibration down to an acceptable, though still noticeable, level.

The principles involved in getting more power from an engine of a given size are exactly the same as those used in "hopping up" an automobile engine, and applying those principles leads to exactly the same problems. The higher-powered engine must deliver more push per stroke, more strokes per minute (with at least equal push per stroke) or both together. For more push per stroke, higher pressure, called BMEP (for Brake Mean Effective Pressure), must be generated by the combustion of the fuel. Doing this calls for a higher compression ratio, for better "breathing," or for supercharging, and in each case the antiknock qualities of the available fuel set a limit to how far the process can be carried. The main limitation of an engine's ability to deliver power at high revolutions is its ability to breathe. Good breathing calls for large valves, for a long duration of valve opening, and for manifold passages and carburetors of ample size: without these, the engine takes in less and less mixture per stroke as it speeds up, resulting in a drop-off in BMEP.

Both higher rpm (revolutions per minute) and higher BMEP make increasing demands on the engine's structure. Bearing loads go up, there is more heat to be dissipated, valve springs are more apt to break, the valves are more likely to burn, the oil gets hotter, and spark plug life suffers. Considering everything, it is not surprising that many advanced engines went through a long debugging period before their reliability matched that of older types. In addition, higher speed in itself creates certain problems. In a high-speed motor the cams must open the valves before and close them after the pistons reach the ends of their stroke. Inertia of the exhaust and intake gases makes this work well at normal operating speeds, but at cranking and idling speeds this inertia is a negligible factor. As a result, unscavenged exhaust gas mixes with the new charge, the power stroke is weak and uneven, and both ease of starting and ability to idle slowly suffer. I have seen a newly overhauled 1917 Mercedes, an engine that developed its power at only 1400 rpm, idling so slowly that each explosion could be counted. It must have been ticking over at less than 100 rpm, about one-fifth or one-sixth as fast as a modern engine. This slow idle was most useful in bringing an airplane to a stop in the days before brakes. Higher speed also means lower propeller efficiency, leading to the requirement for reduction gearing to allow the propeller to turn slower than the engine. This in turn leads to a whole new set of problems. Some engine makers gained the knack of making good gearsets early in the game; others took a long time to learn. It is relatively easy to build good reduction gears, but for a long time it was impossible to make them light; and if they were heavy, all the advantages of the smaller engine would disappear.

Types and Configurations

Aircraft engines have been built in every shape possible for a piston engine. Inlines are much like the usual automobile motor (V-types can be considered a subtype of inlines). They have multiple crank throws and are relatively long and narrow. To some extent, they are better suited to liquid cooling than to air cooling because only the front cylinder gets an unobstructed airstream. Radials are peculiar to aviation. They have a drumlike crankcase with cylinders radiating from it like spokes. They go well with air cooling, and only one significant radial has been liquid-cooled. Radials have almost always been lighter than comparable inlines, partly because of being air-cooled and partly because their longitudinal compactness helps make them stiff without structural weight. Horizontally opposed engines have their cylinders in pairs on opposite sides of the crankcase, just as on the familiar Volkswagen and Subaru automobile motors. They tend to be lighter than inlines for the same reasons as radials. If an opposed engine had the same number of cylinders as a radial of the same power, it would be heavier, but as a rule opposed engines match radial weight fairly closely by having fewer cylinders. Since 1909 all opposed engines have been air-cooled.

Drag and Air Resistance

The problem of frontal area received no real attention until 1917. For one thing, speeds were not so high that engine bulk mattered much. What misled designers for some time was the fact that Allied fighters were generally faster than German fighters despite the fact that all British and French pursuits used engines that were bulkier than the German inline types. When as nicely streamlined a plane as the Albatros was slower than the boxy S.E. 5, there seemed to be little reason for concern about frontal area: it does not seem to have been realized that the speed of the S.E. 5 was due to lighter weight and to the use of a flatter airfoil than the Germans employed [Fig. 1-2(a) and (b)].[1] Also, a short engine rather than a slim one was favored because shortness allowed weights to be concentrated in the interest of maneuverability.

Two American engines of 1917 were the first to incorporate reduced frontal area among their design objectives. The best known of these was the government's Liberty, a big twelve whose two banks of cylinders were set at the narrow angle of 45°, instead of the usual 60°, to minimize width. This effort was fairly successful. In most Liberty installations the height and width of the radiator matched those of the engine and the fuselage that resulted was narrow. Nevertheless there was just so much that could be done as long as cylinders of large bore and long stroke had to be used to obtain the desired power. The real possibilities of improvement lay in the direction of smaller displacement and higher speed, and this was the philosophy that lay behind the K-12, designed by Kirkham for the Curtiss firm. While the Liberty, running at 1850 rpm, needed 1649 cu in. (27 L) to develop 400 hp, the K-12 got the same output from 1145 cu in. (18.76 L) by running at 2500 rpm.

In racing aircraft equipped with smooth "surface" radiators, the K-12 and its successors set numerous speed records. But in everyday use, the gains from reduced frontal area were largely negated by the continuing problem of radiator drag until the introduction of ethylene glycol (Prestone) in the late twenties allowed smaller radiators to be used. In Europe, there was even a trend toward a very bulky kind of engine, called the "W-type," with three banks of cylinders, and airplanes with these engines performed as well as any others.

We had a special problem with radiator area in the United States since our summer weather is warmer than Europe's. European planes that per-

---

[1]The Spad, also faster than German types, had a somewhat more highly cambered airfoil but had a higher wing loading.

(a)

(b)

FIG. 1-2 Why streamlining was long underrated as a factor in performance—two rivals of 1917, one streamlined and one boxy, with the boxy one having slightly better performance: (a) British S.E.5, 1917—Hisso engine, preferably genuine but more commonly license-built as Wolseley Viper or Sunbeam; (b) German Albatros D-III, 1917—180-hp Mercedes.

formed well at home, notably the Fokker D-VII, overheated here. The performance of many American planes, forced to employ larger radiators, suffered by comparison with similar European craft. This led to an interesting result in North Africa during World War II. Spitfires and Messerschmitts used in desert heat had to be "tropicalized" with larger radiators, and performance fell off to the point where P-40s could fight Messerschmitts with some chance of survival, something they could never do in northern Europe.

Prestone cooling might have led to complete displacement of the bulky radials in applications calling for high performance had not the Townend Ring and the NACA cowl come along at about the same time. The Townend Ring, a British invention, was a narrow cowling of airfoil section that was placed around radial engines. Often called a "drag ring" in the United States, it did not reduce frontal area but did confine the turbulence of the cooling air to the engine's diameter. It was also said to generate a forward thrust component from the expansion of the air as it was warmed in passing over the cylinders. Installed in an airplane whose fuselage was narrower than the engine's diameter, as was common before the era of cowled radials, it might add 10 to 15 mph (16 to 24 km/hr) to the top speed.

The American NACA cowl was named after NACA (the National Advisory Committee for Aeronautics), forerunner of NASA (the National Aeronautics and Space Administration) and the organization that gave the world most of its significant aerodynamic advances during the 1925–1940 period. It was a more sophisticated device than the Townend Ring, being longer and internally contoured to slightly pressurize the cooling air. Not as adaptable to existing aircraft as the Townend Ring because it required a bulkier fuselage to blend with its lines, it probably raised speeds by 20 to 25 mph (32 to 40 km/hr) on suitable planes [Fig. 1-3(a) and (b)]. A beneficial side effect was that its confined airflow allowed internal baffling to be provided, leading to better cooling for two-row radial engines.[2]

---

[2]These were, in effect, two radials built one behind the other. While used early in Europe, we avoided them until the NACA cowl came into use.

The final American development in cowls for radials came in 1934 with the introduction of small adjustable flaps (which the British called "gills") at the rear to control the area of the cooling-air outlet. Besides further reducing drag, cowl flaps allowed the flow of cooling air to be restricted during cruising conditions, when less was needed. Previously, without adjustable flow, an engine that was adequately cooled during takeoff and climb was overcooled during cruise. Water-cooled engines had never had this problem because radiator flaps for temperature control had been in use since before 1914.

While American and British sophistication in cowl design seldom went beyond the provision of cowl flaps, the Germans and the Japanese took it a step further. By mounting a multibladed fan on either the propeller shaft or a faster-running shaft coaxial with it, such aircraft as the Focke-Wulf 190 and the Kawanishi "Rex" were able to achieve the ultimate in low-drag cowling design for radial engines (Fig. 1-4). The Japanese may have been oversold on this idea, as some of their installations tended to overheat in the warmer combat zones, but at its best it was an admirable system.

Today there is a good deal of evidence to suggest that engine bulk has been exaggerated as a handicap to high performance. Anything that has ever been done by a slim engine has ultimately been matched by a well-cowled bulky one. Indeed, until recently the world's unlimited speed record was held by a radial-engined Grumman Bearcat. But high speed has always been somewhat easier to attain behind a slim engine; time and again, it has taken considerable time and work before radial-engined planes were able to match the performance of inline-engined craft. The final verdict is that speed can be had from a bulky engine but is easier to get from an engine with less frontal area.

Altitude Performance

Holding power to high altitudes has been almost entirely due to *supercharging,* definable as the pressurization of the intake by a blower. While most superchargers have been gear-driven at several times the engine speed, the more effective type is the turbosupercharger, in which the escaping exhaust gases drive a turbine to power the blower.

American sources credit Dr. Sanford Moss, of General Electric, with inventing the turbosupercharger just after World War I; it is true that we made the first successful application—on a Liberty-powered D.H. 4—but actually Rateau, in France, built the first one in 1913. The nice thing about the turbosupercharger is that it gives the most boost at high altitudes, where it is most needed. This is because the same lowered atmospheric pressure that makes supercharging necessary also reduces back pressure against the exhaust, and so the turbine delivers more power higher up. Unfortunately, the virtues of the turbosupercharger could not be fully exploited during World War II because the heat to which the turbine was subjected meant that it had to be made of high-tungsten alloy steel, and tungsten was in short supply.

Some manufacturers, notably Pratt and Whitney, Rolls-Royce, and Daimler-Benz, got excellent performance from gear-driven superchargers. On the other hand, our rugged Allison was relegated to a somewhat secondary position because of its maker's lack of success in developing a really first-class geared supercharger. Turbosupercharged, as it was in the P-38, the Allison performed well at high altitudes, but in the P-39 and the P-40 it was strictly a medium-altitude engine. BMW (Bayrische Motoren Werke) had a similar difficulty, with the result that the Focke-Wulf 190 was not the effective high-altitude fighter that the Daimler-Benz-powered Messerschmitt 109 was.

Durability

The pursuit of reliability has led to a valuable by-product in the form of increased durability. Durability is measurable as TBO (Time Between Overhauls), and TBO today is at least 20 times what it was in 1914. Most, though not all, of this gain has come from the same improvements in metallurgy, lubrication, detail design, and manufacturing accuracy that were first introduced to promote better reliability. In addition, certain changes, notably hard-chrome-plated piston rings and cylinder walls and nitrided cylinders, have been made purely in the interest of long life. There is some overlap between reliability and durability. In World War I,

FIG. 1-3(a) Boeing P-26 "Peashooter" as seen from the ground. Note that the fuselage is still narrow, as with an uncowled engine, giving the pilot a degree of downward visibility. P & W Wasp, 550 hp. (*Courtesy of the author, taken at NASM-Silver Hill.*)

most engines were reliable enough for the short missions then common but had to be operated at reduced power when, as on the Zeppelin raids, several hours of continuous running were called for. In contrast to this, most engines built since perhaps 1940 can be run for hours at "high cruise" with little increase in the likelihood of failure—though at some cost in reduced TBO.

Fuel Consumption

In all the areas so far mentioned, recent engines have been greatly superior to their ancestors. Very little progress, though, has been made in improving fuel economy. The relative operating economy of different engines can be evaluated by comparisons in terms of specific fuel consumption of X lb (Y kg) of fuel per horsepower per hour. [For the sake of brevity this will be shortened to "specific consumption of X lb (Y kg)" when reference is made to this characteristic in the rest of the book.] The best engines of 1915 had lower specific consumption

FIG. 1-3(*b*)   Howard Hughes Racer, world's fastest landplane in the mid-thirties. Note that the cowling diameter is brought in to meet a fuselage that is still fairly slim, which was common before 1937. P & W R-1535 Twin Wasp Special, 14 cylinders. (*Courtesy of NASM-Smithsonian.*)

FIG. 1-4   Kawanishi N1K "Rex," Mitsubishi Kasei—14 cylinders, 1530 hp, fan-cooled. Few aircraft ever had such small air intakes for such a powerful engine. (*Courtesy of the author, taken at Willow Grove NAS, Pennsylvania.*)

than today's popular lightplane powerplants. This may be a price that has to be paid for the high specific output of modern engines. The kind of early-open, late-close valve timing that is needed for high rpm at takeoff is probably not optimum for economy at cruising power. In practice, the situation is worsened by the fact that a fast-turning propeller is less efficient than a slower one. As discussed in Chap. 3, a modern engine would have to deliver 113 hp to produce the same thrust as a 90-hp OX-5. Thus the modern flier not only has to buy more fuel per horsepower but also has to use more horses to do the same job. This burden is reduced by variable-pitch propellers and by the better streamlining of modern planes, but there remains much room for improvement. Whether improvement will take place is doubtful today, because benevolent governments have raised the cost of certificating new designs to such impossible heights that a manufacturer experiments at considerable risk.

The Piston Engine Today

Large piston engines are a thing of the past. The small air-cooled, horizontally opposed type of engine, in 4- and 6-cylinder form, reigns supreme in all power ranges below those of the smaller turbines. The piston engine, in this surviving form, is enjoying a healthy old age. Despite the seeming effort of governments to regulate private flying out of existence, there are more piston-engined planes flying today than there ever were in the golden age of the big piston engine. The smaller engines will go on for at least a while; but the story of the big piston engines can be told as a matter of past history.

chapter
TWO

The early birds.

The Gasoline Engine at the Dawn of Powered Flight

Aeronautical knowledge at the turn of the century had reached a level that brought powered flight within reach if a suitable power source could be found. The gasoline engine, largely as a result of the work of Comte Albert DeDion, was likewise approaching the efficiency needed to power a human-carrying plane.

As the mentor of the talented *mécaniciens* Trepardoux and Bouton, the Count had been instrumental in developing the first high-speed gasoline engine, in 1895. Up to then the dominant automobile engine had been the 750-rpm Benz; DeDion's engines ran at 1800 and had the added advantage of being air-cooled. Higher speed meant more power from an engine of a given size and weight, and air cooling lowered the weight still further.

Timed ignition was the innovation that made the high-speed engine possible. The usual ignition system of the mid-1890s had employed a ceramic tube projecting into the combustion chamber. Preheated by a torch and kept hot by the running of the engine, it fired the mixture reliably, but timing for power and efficiency was impossible. Instead of using this, DeDion used electrical ignition, first by contacts within the cylinder that were opened and closed by a cam mechanism, producing a spark as they broke apart,[1] and, later, by spark plugs. Since this action was mechanically controlled, it could be made to take place at whatever point in the engine's cycle was most advantageous. Speeds as high as 4000 rpm were attained with these little wonders in tests, and users found them to be, if anything, more reliable than the older type.

At 5 hp, DeDion's engines were too small to power an airplane and apparently were not amenable to a scaling-up process. The young Brazilian Alberto Santos-Dumont made the first-ever goal-and-return flight with a DeDion engine, but this was in a dirigible balloon. When Santos-Dumont later turned to heavier-than-air flight, he used other makers' engines. One might think that power enough for flight could have been gotten out of the DeDion design by developing a multicylinder version,

---

[1]This "make-and-break" system had been invented earlier but was not used to make higher rotational speeds possible before DeDion's work.

11

but this was never done. Perhaps the Count, bereft by 1900 of the assistance of Trepardoux and Bouton, was fully occupied with his automobiles, which were selling as fast as he could build them.

American Engines before 1913

The first engine actually capable of flight was the American Manly-Baltzer. It owed little to DeDion's work, its builders having instead gone to extraordinary lengths in achieving lightness in a large low-speed type. The result of these efforts was a masterpiece, but, like many others, it was never duplicated. The large ultralight engine proved to be a blind alley, and the future lay with the DeDion concept.

Charles M. Manly, then a recent Cornell graduate, was assistant to Prof. Samuel P. Langley of the Smithsonian Institution, the best-known and best-financed aerial experimenter in the world at the time. A large steam-powered model of Langley's design had made the longest free flight on record, and he had begun work on a full-size human-carrying version. As early as 1898, only the lack of a suitable engine seemed to stand between Langley and success. However, overcoming this lack proved to be a formidable task. No engine builder in Europe or America could supply a light enough engine of sufficient power, and so Manly reluctantly took on the task of designing one. Building on the work of Stephen M. Baltzer of New York, he turned out a 5-cylinder radial engine of 540 cu in. (8.85 L) that gave a reliable 52 hp at 950 rpm. (Baltzer's work is more fully described in Chap. 4.) Despite being water-cooled, it weighed only 136 lb (62 kg) exclusive of water and radiator. To make such a big engine so light, Manly went to remarkable extremes. The cylinders, made with a steel shell for strength and a cast-iron liner for good lubrication, had a total wall thickness of only ⅛ in. (3.06 mm). The sheer workmanship required to machine closed-end steel shells 5 in. (127 mm) in diameter, 9 in. (229 mm) long, and 1/16 in. (1.6 mm) thick on 1900 equipment is staggering. On top of this, Manly and his staff had to make equally thin cast-iron sleeves accurately enough to be a shrink fit in these shells. The water jackets were made of sheet steel 0.020 in. (0.5 mm) thick, brazed to the cylinders. This, too, was a formidable

job; it is extremely hard to bring two pieces of metal of very different thicknesses to red heat without having the lighter piece burn before the heavier one gets up to temperature.

To connect the five pistons to one crank, Manly devised a combination master rod system and slipper bearing system. The top cylinder carried a master rod whose lower end had a bronze bearing inside and whose outer diameter was finished smoothly and accurately to allow the other rods' slipper bearings to slide on it (Fig. 2-1). Manly had to create the accessories as well as the engine itself. He designed and built the carburetor, the ignition system (including even the spark plugs), the water pump, and the propeller drive arrangements. The failure of Langley's efforts is, most emphatically, not traceable to any deficiencies in its powerplant; in fact, the Langley "Aerodrome" actually did fly with its original engine. In 1913, Glenn Curtiss refurbished it and, with some modifications, got it off the water under its own power for some short hops.

This engine's story has been told in two monographs published by the Smithsonian Institution. In the first of these, Manly marred the record of his fine achievement by writing a self-glorifying account that denied Baltzer credit for his essential early contribution. Following complaints from Baltzer's heirs, Robert Meyer, the curator of aircraft propulsion, issued another monograph[2] that set the record straight while still giving Manly the credit he deserved. This very readable work is heartily recommended to any reader interested in the full details. The best measure of Manly's—and Baltzer's—achievement is that it took 16 years before another engine was built with a power-to-weight ratio equaling that of the Langley radial.

The Wright brothers' aerodynamic and structural design was sufficiently advanced beyond Langley's to allow the use of a much less sophisticated powerplant. The original Wright engine, built to the brothers' design by Charles E. Taylor, was an inline four that operated lying on its side. The inline configuration is inherently heavier than the radial, but this disadvantage was minimized by casting the block in aluminum, then a very new metal. It gave 16 hp when first started, then settled

[2]No. 6, *Smithsonian Annals of Flight,* Washington, 1971.

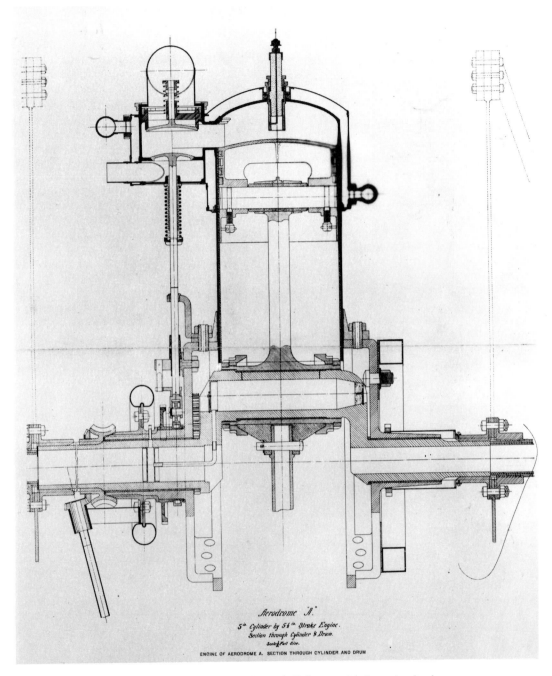

Aerodrome A.
5" Cylinder by 5¼" Stroke Engine.
Section through Cylinder & Drum.
Scale ⅜ Full Size.
ENGINE OF AERODROME A. SECTION THROUGH CYLINDER AND DRUM

FIG. 2-1 Cross section of Langley's Manly-Baltzer engine. It can be clearly seen
how the bronze slippers of the connecting rods of cylinders nos. 2 through 5
bore on the smoothly finished outside of the master rod. The action of the
internally coned nuts that retained the slippers and allowed wear to be taken up
is also clear. What is not at all clear is how the master rod was assembled onto
the crank. *(Courtesy of NASM-Smithsonian.)*

13

down to a steady 12 at 1090 rpm. It displaced 200 cu in. (3.27 L) and weighed 179 lb (81.4 kg). At 14.9 lb/hp (6.77 kg/hp), its weight-to-power ratio was far below that of the Manly engine, and, despite its slightly higher speed, its specific output, at 1 hp per 16.67 cu in. (272 cc) was below the Manly's 1 hp per 13.2 cu in. (21.55 cc).

The valves were in pockets atop the cylinders, the exhaust below, and the intake on top as it lay on its side. The photo (Fig. 2-2) makes the exhaust arrangements reasonably clear; the valves were actuated by rockers from a chain-driven camshaft, and the exhaust gas passed out through the slots in the valve cages, there being no exhaust manifold. On the intake side it had neither carburetor nor manifold nor valve mechanism. The Wright engine shared with the Manly and most other four-stroke motors[3] of the period the use of automatic intake valves. That is, as the piston descended on the intake stroke, the resulting suction overcame the resistance of a light spring to pull

[3]The usual four cycles have been described as: suck, squish, pop, ptui.

the valve open. In place of a carburetor, it used a system that has been described as *fuel injection,* although the term is not really justified. Fuel was simply allowed to drip into the intake ports, where enough was picked up by the airstream to keep the engine running. Since the intake was contiguous to the exhaust and there was no water jacket around either, the area was hot enough to promote adequate evaporation of the gasoline. It had no throttle, such control as there was coming from advancing or retarding the spark, which operated on the make-and-break principle. In operation, no throttling down was called for; even in the Wright's light and efficient airframe, the engine had to run wide open at all times to sustain flight.

The Wright engine, unlike the Manly, sired numerous offspring. Throttle control and a normal carburetor were used on later models, and before all traces of the original design had disappeared in the course of evolution, Wright 4-cylinder motors were delivering 30 hp from a weight of 180 lb (82 kg). Sixes, giving up to 60 hp, were also built before the First World War. Wright became Dayton-Wright, Wright-Martin, Wright Aeronautical, and

FIG. 2-2   The original Wright Brothers' engine—the first ever to carry a man into the air in a heavier-than-air machine. While hardly an advanced design, even for 1903, it did the job. (*Courtesy of NASM-Smithsonian.*)

finally Curtiss-Wright, but it never left the engine business. When the production of Wright radials came to an end in the 1960s, long after the firm had stopped building airframes, Wright was the oldest continuous producer of piston aircraft engines in the world.[4]

After the Wrights, Glenn Curtiss was America's best-known aviation pioneer. He came originally to aviation as an engine builder. His reputation as a maker of motorcycle engines led Capt. Thomas Baldwin to come to Curtiss for a motor for his dirigible. This engine was a V-twin with more serious-looking cooling fins than most contemporary air-cooled engines and with auxiliary exhaust ports drilled around the cylinder to release some of the exhaust gas when the piston reached bottom dead center. Several early engines used this device, but Curtiss added a refinement in the form of a perforated ring that could be rotated to vary the effective area of the auxiliary ports.

Curtiss became the fastest man in the world by traveling at over 120 mph (200 km/hr) on a motorcycle powered by a V-8 that was in effect four of the Baldwin-type V-twins. He was then invited to join the AEA (Alexander Graham Bell's Aerial Experiment Association). Though he came to the AEA basically to be its powerplant specialist, he also built the most successful of the group's three aircraft.

Much of the casting and machining work for Curtiss's early engines was done by Charles B. Kirkham, whose Savona, New York, shop was close to Curtiss's Hammondsport home. As many readers have probably seen a television special on the Wrights that portrayed Curtiss as a thief of others' ideas, and since Kirkham later had a distinguished career in engine design, it may be suspected that Kirkham was behind the motors that bore Curtiss's name. The facts are otherwise. If anything, Kirkham's tutelage under Curtiss provided much of the inspiration and directing force behind Kirkham's later career. Glenn Curtiss never had to steal anyone's ideas. His own, as with the aileron, were usually better.

Curtiss built mostly V-8s. At first they were air-cooled and, for V-8s, surprisingly small for the period (Fig. 2-3). His first was called the B-8 and

[4]Rentschler's 1919 reorganization of Wright-Martin into Wright Aeronautical is counted as continuous operation.

FIG. 2-3   Curtiss B-8, 40 hp, about 1907. *(Courtesy of the author, taken at NASM Hall of Early Flight.)*

was essentially the same as the record-setting motorcycle engine; it gave 40 hp. It was followed by the water-cooled E-4, a four with larger cylinders and an output of 50 hp; then eight of the same cylinders went into the 100-hp E-8. This was the powerplant that amazed the Europeans by setting a world's record of 46.5 mph (75 km/hr) at the Rheims meet and was the last of Curtiss's air-cooled motors. The first of the water-cooled V-8s had been built even before the E-8 (Fig. 2-4), and the 90-hp OX-5 soon developed from it. As this famous (perhaps infamous) machine was more properly a wartime powerplant, it is dealt with in Chap. 4.

The mystery man among early engine builders was Gustave Whitehead, *geboren* Weisskopf. This German immigrant built both airplanes and engines in Bridgeport, Connecticut, before 1900; his degree of success is a matter of controversy and speculation. I tend to accept the opinion of Mr. Harvey H. Lippincott, the very knowing Pratt and Whitney archivist, that Whitehead probably got off the ground for short, uncontrolled hops as early as 1901 with a "carbide" engine for power. Such an engine would have used water dripping on calcium carbide to generate acetylene gas, which either burned or expanded in the engine's cylinders. After 1900 Whitehead built gasoline engines and sold a

FIG. 2-4   Early Curtiss water-cooled V-8, model V2-C-1C of 1914. *(Courtesy of the author, taken at Glenn Curtiss Museum, Hammondsport, New York.)*

few. It is probable that other experimenters did more flying with Whitehead's engines than Whitehead himself ever did.

After Curtiss, the best-known early U.S. engines were the San Francisco–built Hall-Scotts. Although it made some V-8s (Fig. 2-5), the company was probably better known for straight fours and sixes. These engines were developed as early

FIG. 2-5   Early Hall-Scott V-8. *(Courtesy of the author, taken at NASM Hall of Early Flight.)*

as 1907 and employed separate cylinders and overhead camshafts, much as did the German engines, although the Hall-Scotts were generally lighter and less reliable than the German engines. The Navy's largest aviation facilities before the war were at North Beach, near San Diego; as the closest source of engines to that base, Hall-Scott naturally enjoyed advantages with regard to liaison and servicing. In those innocent days before it learned the art of separating authority from responsibility, the Navy trusted its ship and station commanders with considerable discretion, and so Hall-Scott sold a fair number of engines to the Navy. They went into tractor biplanes built by Glenn Martin, the Navy having dropped its Curtiss pushers after a number of pilots had been crushed when the rear-mounted engines broke loose in crashes. While the Hall-Scotts were not notably more reliable than Curtiss engines, survival rates were enhanced by the fact that in a crash, the engine, and not the pilot, hit ground first.

Two well-known American pioneer engines used the two-stroke principle. The great appeal of the two-stroke design lay in the smaller number of moving parts, as compared with four-stroke engines, and the resulting decrease in the likelihood of something failing in service. In particular, there are no valve springs to break and no lubrication system to fail. Since Palmer Brothers (The Cos Cob, Connecticut, marine engine builders) had introduced the practice in 1903, it had become usual to lubricate two-strokes by mixing oil with the fuel, just as we do with outboard motors today. Thus, as long as a two-stroke is getting fuel, it is also being lubricated. Two-strokes do have the disadvantage of excessive fuel consumption, often having a specific consumption 60 percent higher than comparable four-strokes, and the type largely died out for aviation use when the reliability of four-strokes reached acceptable levels (Fig. 2-6).

The first of these two-stroke specialists was Elbridge, a maker of marine engines in Rochester, New York, who got into airplane engines almost by accident. Its runabout powerplants were quite light, and several were sold to airplane builders between 1908 and 1910. As a result, the firm began to advertise its ''Featherweight'' engines as suitable for aviation use in 1910 (Fig. 2-7). The ''Featherweight'' line comprised a 30/45 hp, 150-

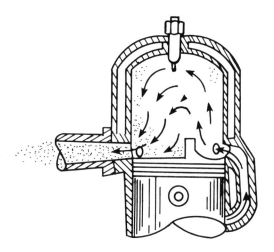

FIG. 2-6 Scavenging in a two-stroke. Why two-strokes use a lot of fuel. The incoming fuel (arrows) is supposed to push the exhaust out ahead of it (dots). This is called "scavenging" and it is imperfectly done, since some of the fuel escapes and/or some of the exhaust stays to contaminate the new charge.

lb (69-kg) three, a 40/60 hp, 200-lb (89-kg) four, and a 60/90 hp, 250-lb (118-kg) six. All three were rated at 1400 rpm, fairly fast for the time. This choice of number of cylinders was common with all kinds of two-strokes. Since each cylinder of a two-stroke is somewhat more of a self-contained unit than cylinders in a four-stroke, it is relatively easy to use the same cylinder assemblies for a whole range of engines with however many cylinders the buyer wants.

Encouraged by good sales of the "Featherweights," Elbridge designed an engine specifically for aircraft in 1911, when, unfortunately, the day of the two-stroke was almost over. This was its "Aero Special," a 50/60 hp four whose most notable feature was its high speed of 2000 rpm. No propeller of the time could have taken this speed had the engine been larger, but a 60-hp 2000-rpm engine only needs about a 75-in. (2000-mm) propeller, and so a suitable combination was possible.

The other well-known American two-stroke was the Roberts, made in Sandusky, Ohio, as late as 1919. There is one in the Bradley Air Museum, donated by the late Harry Ford of Bridgeport, Connecticut. Ford had used it in a Curtiss-type

"hydroplane" that he built himself with assistance from the sometimes-maligned Glenn Curtiss. It looks like a light marine engine of the period except for having aluminum cylinders instead of cast-iron ones. The later Roberts engines had an arrangement of auxiliary intake ports in an effort to hold down the fuel consumption, but by 1914 nothing could save the two-stroke engine (Fig. 2-8).

Lowest priced of early engines was the Detroit Aero of 1909. This was one of several engines designed by Fred Weinberg with the ostensible aim of bringing prices down to a level more affordable to larger numbers of enthusiasts than were expensive conventional engines. The Detroit was a rather large [237.5-cu.-in. (3.89-L)] air-cooled opposed twin, rated at 25/30 hp at 1500 rpm. A picture of this engine in Glenn D. Angle's 1921 *Aero Engine Cyclopedia* shows it with a very crude propeller consisting of two flat wooden blades bolted to a hub made of a twisted steel plate. It was claimed that over 1000 were sold, and perhaps as many as 500 actually were, but nothing like this number were used in airplanes that were completed and flown. This engine has somewhat the aura of an

FIG. 2-7 Elbridge "Featherweight," 30/45 hp, about 1910. Note the way the intake manifold gives the same good fuel distribution that a dual carburetor would, without the added complication. I once had a Tuttle 3-cylinder two-stroke marine engine that was made this way. *(Courtesy of the author, taken at NASM Hall of Early Flight.)*

FIG. 2-8 Four-cylinder Roberts model 4X. Note the lack of provision for an exhaust manifold. It is most important to avoid back pressure in a two-stroke; in the early days no one knew how to design an exhaust manifold for zero back pressure. This engine must have been just a little noisy! *(Courtesy of the author, taken at Glenn Curtiss Museum, Hammondsport, New York.)*

attempt to exploit the hopes of ill-informed and underfinanced backyard builders.

Meanwhile, Charles B. Kirkham had started on the course that eventually led to the great D-12 and, ultimately, to the Merlin (see Chap. 5). After building a few engines under his own name at Savona (Fig. 2-9), he joined the Aeromarine firm, at Keyport, New Jersey, as chief engineer. By 1910 or so, advances in automobile engine practice had made it possible to seriously consider building aircraft engines along automotive lines. The appearance of the Aeromarines, and of the Sturdevants from Boston, Massachusetts, suggests such an approach. The typical large auto engine of around 1910 had its cylinders cast in pairs, with integral heads, and bolted to the crankcase; the Aeromarines and Sturdevants looked a good deal like such car motors. The main difference was the use of aluminum instead of iron for the cylinders and crankcase castings. Sturdevants found some buyers among independent builders, and Aeromarines powered the efficient flying boats that gave the firm its name (Fig. 2-10).

There is one other American engine which belongs to the war era if chronology alone is considered but which may well be thought of as the last product of the early experimental period:

the Duesenberg. Built in Elizabeth, New Jersey, it used horizontal valves operated by rocker arms that were so long that they were called "walking beams" after the prominent abovedeck connecting levers of paddle-wheel steamboats. Although this design seems somewhat pointless in the light of modern knowledge, Duesenberg racing cars were America's best for a while, and the idea of developing an aircraft version was far from an absurd one. The firm made some very large engines, including a V-16, but there were few if any takers. Late in the war, the Duesenberg plant was used to produce the Bugatti-King engines.

It is possible that this incomplete list of early types will leave the reader with the impression that the United States was a beehive of aeronautical activity during the years before the war. Nothing could be further from the truth. Though 500 Detroit Aeros may have been built and sold, it is probable that this is more than the total number of airplanes of all types made and flown in the United States before 1913. Only Wright, Curtiss, Martin, and Aeromarine were serious commercial factories, and their production was on the smallest of scales. The rest of the comparatively few airplanes in the country were imports, largely Bleriots, or they were the work of amateurs and semi-custom builders. Flying got a lot of publicity, but it was a case of a good deal of smoke coming from a very small fire.

Europe: The Continent

Europe was rather more active in aviation after 1908, and much the same sort of proliferation of engine types took place. Engine manufacture there was on much the same semi-custom basis as here. However, the lower cost of skilled labor in Europe tended to make this sort of production a little less expensive than in the United States, especially as there was nothing like our low-cost mass-produced automobile engine (Oldsmobile, Ford, Hupmobile) to which the cost of a semi-custom job could be compared.

The best-known early European aircraft engines were the Antoinettes, built by Levasseur. A 24-hp Antoinette was used by Santos-Dumont in 1906 for Europe's first significant flight, and a 50-hp Antoinette V-8 powered the Voisin in which

FIG. 2-9 Kirkham B-6 of 1915. A neat enough motor of its kind but a far cry from the K-12 and D-12 of only a year or so later. *(Courtesy of the author, taken at NASM Hall of Early Flight.)*

Farman, in 1907, made the Continent's first fully controlled flight.

The Antoinettes used aluminum for the crankcase and cylinder heads and used electro-formed (made by plating onto a form that was

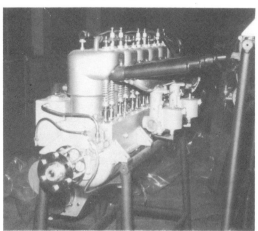

FIG. 2-10 Sturdevant, probably a model 5A of 1917. A decidedly obsolete design if it was really produced that late. *(Courtesy of the author, taken at NASM Hall of Early Flight.)*

subsequently removed) copper water jackets. They employed the same "F-head" configuration, with automatic intake valves, as the Manly motor and had a fuel injection system more advanced than the Wrights'. Cooling was by evaporation, but I am unable to determine whether this was a total-loss system or if the steam was condensed and reused; probably it was the former, because no condenser is in evidence on the best-known Antoinette-powered planes. They used spark plugs rather than make-and-break ignition. They had a large number of exposed parts, resulting in a complicated appearance, but nevertheless gave an impression of being beautifully built.

Antoinettes were said to have been easy to control in the air. There is room for skepticism about this because, since there was no way to throttle the airflow, the only means of control would have been by varying the mixture strength or by retarding the spark. In my opinion, had the brutally simple injection systems of the Wright and Antoinette engines been satisfactory, they would have been widely used, because the early carburetors that constituted the only alternative were a continual headache.

Although the Antoinettes were widely used before 1909, they soon faded from the scene. They required constant and devoted attention and must have been extremely expensive. V-8s were complex enough, but Levasseur also made V-16s and V-32s; it soon became unnecessary to have recourse to such elaboration when more power was wanted. Probably the short life of France's other V-8, the E.N.V., also had something to do with cost.

While the Antoinettes and E.N.V.s were the sole French engines at the high end of the power range, there were numerous smaller and cheaper motors. Of these, the Duthiel-Chalmers was the first to become well known. A 20-hp air-cooled opposed-twin Duthiel-Chalmers powered Santos-Dumont's famous "Demoiselle," and the later 24-hp water-cooled twin was fairly popular. At least one Duthiel-Chalmers, an 8-hp twin, was exported to the United States, where a Georgian experimenter named Sellers flew successfully with this very small engine. In the Duthiel-Chalmers, the endemic French technical exuberance expressed itself in the use of sleeve valves in the cylinder heads in place of the usual poppet type. This should not be confused with the later Knight and Burt-McCollum sleeve-valve designs, which used the sleeves for

the cylinder wall (see Chap. 7). The Duthiel-Chalmers engines were conventional below the cylinder heads.

Two notable automakers of the time tried their hands at opposed engines. Clement-Bayard had first built a 7-cylinder water-cooled radial that operated in a flat position with the shaft vertical, driving the propeller through bevel gears. Its flat engine was a water-cooled 30-hp twin, introduced in 1910, by which date the interest in such small motors was decreasing (Fig. 2-11).

Darraq also built opposed engines, the first a 24-hp twin of 1909, which is credited by some sources with being the powerplant of the "Demoiselle." This engine was then, in effect, doubled to make a 48-hp flat four—probably the first of its kind. Both Darraqs were water-cooled. As specific output began climbing above 1 hp/8 cu in. (131 cc.), engine designers were finding that the old air-cooling arrangements, with their too-simple cylinder head-cooling fin designs, were no longer adequate, as they had been when less power and heat were being generated.

However, some firms never wavered in their faith in air cooling: the most notable of these was Anzani. Anzani's first engines introduced a new configuration called the "fan-type" that enjoyed a

FIG. 2-11   Clement (or Clement-Bayard) 30-hp twin of 1910. *(Courtesy of NASM-Smithsonian.)*

few years' vogue among French aviators. A fan-type motor looks like a single that has been converted to 3 cylinders by adding lugs to the connecting rod to which are attached link (articulated) rods from the added cylinders [Fig. 2-12 (a) and (b)]. The short-lived REP, named for its builder, Robert Esnault-Pelterie, was another fan-type.

Fan-type engines were inherently rough running, since there is no way to get an even firing order, but their compactness led to their having a fairly good power-to-weight ratio. The Anzani that powered Bleriot's cross-channel flight was typical. It gave 24.5 hp from 206-cu-in. (3.375-L) displacement at 1600 rpm and weighed 145 lb (66 kg), roughly 5.9 lb/hp (2.69 kg/hp). The intake valves were automatic, and the cooling fins went only halfway down the cylinders. Cooling was marginal. On Bleriot's cross-channel flight, an overheating engine nearly put him in the water, and only a fortunate rain shower that cooled off the motor allowed the flight to be completed.

A more lasting contribution by Anzani was the development of the first practical air-cooled "static"[5] radial engine. The first Anzani radials were built in 1910, had 3 cylinders, and were not noticeably advanced over the fan-types in cylinder and head design. A 22.5-hp model displacing 140 cu in. (2.29 L) was little used, but the 35-hp 190-

[5]As opposed to "rotary"; see both the discussion of Gnome rotaries later in this chapter and also Chap. 4.

FIG. 2-12(b) Anzani fan-type, as used by Bleriot for the first crossing of the English Channel by an airplane. Three cylinders, 206 cu in. (3.38 L), 145 lb (66 kg), 24.5 hp at 1600 rpm. The springs and stems of the automatic intake valves are inside the intake pipes. *(Courtesy of NASM-Smithsonian.)*

cu-in. (3.12-L) model proved to be a classic powerplant for ultralight planes, continuing in at least occasional use for 20 years. Specimens that still exist today are most uncertain in operation. I suspect that this is due to the gasoline. Early gas was far more volatile than anything available today, especially aviation grades, whose volatility is deliberately kept low to minimize vapor lock and other troubles at working altitudes.

The "lower end" of the Anzani radials was interesting in that the master rod system of the fan-types was eliminated in favor of slipper bearings somewhat like Manly's. This was probably done in the interest of smoothness. It can be seen from Fig. 2-12 that the link rods are aligned with the centerline of the crankpin at only one position of the crank. This means that no two cylinders of a radial using the link rod system will have the same pattern of acceleration and deceleration as the crankshaft turns. The fewer cylinders an engine has, the more noticeable the resulting vibration will be, and the slipper bearing system is probably the best way to build a 3-cylinder radial, but its practicality decreases as more cylinders are used. In the end the master-and-link layout became standard for all radials.

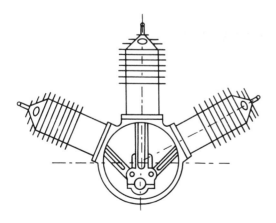

FIG. 2-12(a) Fan-type engine cut away to show master and link rods.

Before this evolution took place, a different system was successfully used between 1912 and 1919. Known as the "Canton-Unne Patent," it was employed by Salmson for a long line of first-class water-cooled radial engines. The Salmson engines themselves are described in Chap. 6; the Canton-Unne system used a cage, bearing on the crankpin, carrying bushings for link rods from every cylinder and geared so as to stay vertical as the crankshaft turned.

The use of automatic intake valves began to die out in France after 1909; only Anzani persisted with them. That they lasted this long is strange, considering that automobile engines had discarded them at least 5 years earlier. There are three things wrong with automatic intake valves: they make for a less powerful engine, even at sea level; they cause an exaggerated drop-off in power with altitude; and they are excessively prone to sticking. The power loss at sea level comes from short duration of valve opening. The valve does not open until the descending piston has pulled a vacuum high enough to let the ambient air pressure open the valve against the force of the spring, and it closes early for the same reason, all of which adds up to poor breathing and cylinder filling. It gets worse at higher altitudes because the ambient air pressure is lower. In contrast, a cam-operated valve can be made to open and close at whatever point in the cycle the designer decides on. The cam-operated valve, in addition, cannot stick closed and, because the spring can be stronger, is less likely to stick open. The fact is that early gasoline-engine designers were too concerned about easy egress for the exhaust and not concerned enough about good breathing. The Darraq probably represented the transition point in thinking about this; it had cam-operated intake valves, but it also had auxiliary exhaust ports, like the early Curtisses, to reduce back pressure.

Renault, like Anzani, opposed the trend to water cooling with its V-8s, first made in 1909, and its slightly later companion twelves. These, like so many early engines, used the F-head configuration. Recent F-heads have the intake valve on top because there is room to make it large in the interest of good breathing, but Renault put the exhaust on top and the intake below. This allowed the exhaust to be piped away without having it heat the cooling air, as would have been unavoidable had the exhaust pipes passed between the cylinder banks [Fig. 2-13(a) and (b)]. In addition, these engines did not depend on the aircraft's forward speed for cooling airflow. They used, instead, a large fan at the rear of the engine to assure adequate airflow, accepting the fact that some power would be absorbed in driving it.

The Renaults' most interesting feature was that the propeller was driven off the camshaft instead of the crankshaft. The camshaft runs at half engine speed in almost all four-stroke engines, and so a built-in reduction gear is available. All that is necessary is to enlarge the timing gears and beef up the front of the camshaft to take the load. This simple system has been tried many times, but few engines other than the Renaults and the later Renault-based RAF (Royal Aircraft Factory) engines have used it successfully. The trouble is that engines need something to smooth out the power impulses, or, more technically, the torque variation that occurs during each power stroke. The propeller performs this function if it is attached to the crankshaft, but, if it is not, the unevenness is transmitted through the gears. The gears thus must be very

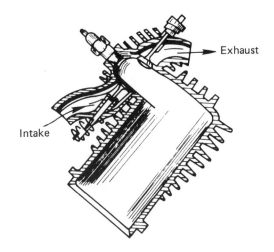

FIG. 2-13(a) Early Renault V-8 cylinder. Material was cast iron. Note that the intake-valve spring was coil but exhaust was "volute."

FIG. 2-13(*b*)   Renault V-8, 80 hp. This particular engine is of 1916 manufacture but is the same as the firm's pre-war engines. The cylinder hold-downs are interesting. The second cylinder's rocker arm and pushrod are missing. Note the very large blower housing. (*Courtesy of NASM-Smithsonian.*)

strong if they have a simple relationship like 2 to 1. Odd ratios like 20 to 39 or 20 to 41 work better because the same tooth doesn't get the same load at each revolution. The camshaft drive worked on the Renaults because they were very "soft" engines, having low compression ratios and low BMEP. In addition, they were sturdy to the point of being heavy. The first of the V-8s, a 35-hp model, weighed all of 242 lb (110 kg). Renault itself evidently came to believe that air cooling was a blind alley, because all of its wartime development work went into water-cooled engines. The air-cooled types continued in production but were not updated, and their main application after 1915 was in training planes. The fact is that there was no real gain from reduction gears until engines began to be operated at above 1800 rpm. The early Renaults ran at 1600 and none of the air-cooled types at over 1800. Planes using these motors had to have four-bladed propellers to absorb the power at 800 to 900 rpm without resorting to excessive diameter.

The Renault V-8s went to 50 hp in 1910 with the 372-cu-in. (6.1-L) model and to 80 hp in

a 548-cu-in. (9-L) wartime version. The twelves included a 651-cu-in. (10.7-L) 90-hp model, one of 742 cu in. (12.2 L), and an 823-cu-in. (13.5-L) 138-hp model. Apparently they were reliable enough, but they were appallingly inefficient, with a fuel consumption of 0.76 lb (0.345 kg)/hp/hr at a time when some engines used only two-thirds as much and the 1916 Hispano-Suiza was getting 50 percent more power from the same displacement.

Of all the early French engines, the one that made the greatest impression on the contemporary aviation world was the Gnome rotary (Fig. 2-14). It was not rotary in the Wankel sense but was a piston engine, of radial form, that rotated around a fixed crankshaft. While a 5-cylinder model was built in 1908, the classic early Gnome was the 50-hp seven of 1909. It ran at 1150 rpm and weighed only 165 lb (75 kg), giving the excellent weight-to-power ratio of 3.3 lb/hp (1.5 kg/hp). The Gnomes worked well enough to be exported to every country that built airplanes and to become the ancestors of a number of other more or less imitative rotaries.

FIG. 2-14   The first (literally, this engine carries serial number 0) Gnome 50-hp 7-cylinder rotary, 1909. *(Courtesy of NASM-Smithsonian.)*

As these engines constitute a distinct and unique type, Chap. 4 is devoted to their story.

Moving to consider the rest of the Continent, we find that it lagged behind France in engines as in other phases of aviation. The first flight in Denmark was made by one Ellehammer, using a radial engine of his own construction of which little is known and which left no descendants. Much the same can be said of the homebuilt two-stroke with which Hans Grade made Germany's first doubtful hop.[6]

By 1910, however, Germany had begun to overtake France as an engine builder. Where engine types had proliferated in France, the Germans (and the Austrians) built only one sort of motor. All the Teutonic aero engines, with the exception of a few Gnome-derived rotaries, were water-cooled inlines with separate, rather than *en bloc,* cylinders. Three of these, the Argus, the Austro-Daimler, and the Mercedes, stand out in the pre-war era. As the direct precursor of a long line, the Austro-Daimler is described in the next chapter, as is the Mercedes. Although a Mercedes won a *Kaiserpreis* in 1911, its full development did not come until after 1914. The Argus, though, was more prominent before the war than during it. Early Arguses powered a majority of the birdlike Taubes that were the most characteristic pre-war German aircraft. By the time the war began, their reliability had become much

better, and they were much favored in Russia until hostilities cut Germany off as a source. In fact, 100-hp Arguses powered Igor Sikorsky's Grand, the world's first four-engine plane: they apparently gave good service.

Russia had only one aircraft engine builder, the Russo-Baltique automobile works in Riga, Latvia. It built imported designs, and even this doubtful source was lost when the Germans took over the Baltic states after the Battle of Tannenberg. In this connection, acquaintance with early engine development makes it impossible to take seriously the Soviet claim that Mozhaisky flew successfully in 1889. While Czarist Russia had more industrial capacity than is sometimes realized, its engineering capabilities were extremely limited. One cannot believe that a country having to import designs for such relatively simple devices as steam locomotives and bolt-action military rifles could have built, before 1900, an engine of adequate power-to-weight ratio for flight. Such technically advanced countries as France and the United States were just barely able to do so, and Britain and Germany could not until they had been shown the way.

The first British engines to become well known were the water-cooled Green inlines. Their angular, inelegant appearance has led to their being called "crude," but they seem to have been decent enough motors. One in particular, the 1913 35-hp four, was good enough to have been put back in production, albeit briefly, in 1919. Green tried to develop some big engines, including a twelve, but inadequate financing stood in its way.

At the opposite pole from the conventional Greens stood the radical NEC. The initials stood for New Engine Company, and new it was. The NEC was a supercharged, fuel-injected two-stroke engine. The idea, valid enough but ahead of its time, was to overcome the scavenging problem by having a blower push in a cleansing surplus of plain air when the ports were exposed, then inject the gasoline after they were closed. It has been said that all efforts to improve the basic two-stroke engine cost more in complexity than they gained in efficiency, and this was undoubtedly true of the NEC.

Between them, these two engines prefigured much of the future history of British engine devel-

---

[6]In the twenties, Grade, by then a prosperous motorcycle builder, reentered the aero motor game with a small two-stroke twin.

opment in that one was sound but conventional while the other was quixotically unrealistic. Yet, in at least one case, British stubbornness in pursuing unlikely ideas paid off handsomely: without the work done on the sleeve valve, the Second World War just might have turned out differently.

Summary

By 1913, then, a great many concepts had been put through the wringer, and the survivors were those of proven viability. In France, the rotaries and the Renault were to be standbys in the coming war. In the United States, the Curtiss line of V-8s had evolved into the OX-5 that was to power the wartime training program—and most private flying up to 1928. The German inlines were to be their wartime mainstay and were to inspire British developments that surpassed their progenitors. The embryonic aluminum-block concept had not yet given birth to the great Hispano, while the pipsqueak Anzani radials hardly prefigured the marvels that were to come; but it was all there, just waiting for technology to catch up with concept.

Data for Engines

Maker's name	Date	Cylinders	Config-uration	Horse-power	Revolutions per minute	Bore and stroke, in. (mm)	Displace-ment, cu in. (L)	Weight, lb (kg)	Remarks
Manly-Baltzer (Langley)	03	5	R, L	52.4	950	5 (127) × 5.5 (140)	540 (8.85)	207.5 (94.3)	Weight is with water
Wright Brothers	03	4	I, L	12–16	1090	4 (102) × 4 (102)	200 (3.28)	179 (81.4)	
Curtiss A-2	NA	2	V, A	7	1500	3.25 (83) × 3.625 (92)	60 (983 cc)	50 (22.7)	Same engine used in record-breaking motorcycle
Curtiss B-8	NA	8	V, A	40	1800	3.625 (92) × 3.25 (83)	268 (4.4)	150 (86)	
Curtiss E-4	NA	4	I, L	50	1500	5 (127) × 5 (127)	393 (6.4)	250 (114)	
Curtiss E-8	NA	8	V, L	100	1500	5 (127) × 5 (127)	785 (12.9)	350 (159)	
Hall-Scott A-2	12	8	V, L	65	1400	4 (102) × 4 (102)	402 (6.6)	260 (118)	A-3 similar but 4 × 5, 80 hp
Hall-Scott A-5	14?	6	I, L	125	1250	5 (127) × 7 (178)	825 (13.5)	525 (239)	First Hall-Scott with overhead camshaft
Elbridge "Featherweight"	10	3	I, L, TS	30–45	1400	4.625 (117) × 4.5 (114)	227 (3.7)	150 (86)	4- and 6-cylinder models similar
Elbridge "Aero Special"	11	4	I, L, TS	50–60	2000	4.625 (117) × 4.5 (114)	200 (3.28)	150 (86)	
Roberts	11	4	I, L, TS	50	1400	4.5 (114) × 5 (127)	318 (5.2)	170 (77)	
Roberts	11	6	I, L, TS	75	1200	5 (127) × 5 (127)	477 (7.8)	243 (110)	
Kirkham BG-6	10	6	I, L	70–90	1680	4.3125 (79) × 5.125 (131)	449 (7.4)	420 (191)	
Detroit Aero	09	2	O, A	25–30	1500	5.5 (140) × 5 (127)	237 (3.9)	NA	
Aeromarine L-6	NA	6	I, L	130–145	1700	4.25 (108) × 6 (152)	553 (9.1)	400 (182)	Available in geared form
Sturdevant D-4	11	4	I, L	48–55	1700	4.5 (114) × 4.5 (114)	268 (4.4)	220 (100)	L-head
Duesenberg A-44	17	4	I, L	125	2100	4.375 (111) × 6 (152)	361 (5.9)	365 (166)	Built 850-hp V-16, 1918
Antoinette	06	8	V, L	32	1400	3.15 (80) × 3.15 (80)	196 (3.2)	93 (42)	Built V-16 also; different references give various hp figures for V-8s

Engine	Year	Cyl.	Type	hp	rpm	Bore × Stroke, in. (mm)	Displacement	Weight	Remarks
E. N. V.	10	8	V, L	39	1700	3.35 (85) × 3.54 (90)	249 (4.1)	150 (68)	Made in Britain also
Duthiel-Chalmers	08	2	O, A	20	NA	NA	NA	NA	
Duthiel-Chalmers	09	2	O, L	24	1200	5 (128) × 5.12 (130)	204 (3.3)	132 (60)	No data on 8-hp model
Clement-Bayard	10	7	R, L	50	1200	3.94 (100) × 4.53 (115)	387 (6.32)	154 (70)	Used for helicopter
Clement-Bayard	10	2	O, L	30	NA	5.25 (133) × 4.375 (111)	181 (3)	110 (50)	Vertical shaft
Darracq	09	2	O, L	24	1500	5.2 (132) × 4.72 (120)	194 (3.2)	121 (55)	Made 4-cylinder version, same cylinders, 48 hp
Anzani	09	3	F, A	24.5	1600	4.13 (105) × 5.12 (130)	206 (3.75)	145 (66)	Power data questionable for these two engines; note
Anzani	10	3	R, A	30	1300	4.13 (105) × 4.72 (120)	190 (3.12)	121 (55)	higher output at lower rpm for smaller engine
Renault	09	8	V, A	35	1400	2.76 (70) × 4.72 (120)	226 (3.7)	242 (110)	
Renault	13	8	V, A	80	1600	4.13 (105) × 5.12 (130)	548 (9.0)	463 (210)	Also made a 50-hp model with 3.6 × 4.8 in. (90 × 120 mm) cylinders
Renault	NA	12	V, A	90	1600	3.54 (90) × 5.51 (140)	651 (10.7)	823 (13.5)	Also made a 130-hp twelve with 4.2 × 5.2 in. (105 × 130 mm) cylinders
Gnome	08	5	Ro, A	34	1300	3.94 (100) × 3.94 (100)	209 (3.4)	132 (60)	
Gnome	08	7	Ro, A	50	1200	4.3 (110) × 4.72 (120)	488 (8.0)	165 (75)	
Argus type II	12	4	I, L	100	1250	5.51 (140) × 5.51 (140)	525 (8.6)	309–345 (140–157)	Conflicting weights published
Argus type II	12	6	I, L	110	1300	4.88 (124) × 5.12 (5.12)	575 (9.4)	430 (195)	
Green C-4	13	4	I, L	32–35	1200	4.13 (105) × 4.73 (120)	253 (4.1)	160 (73)	
Green E-6	NA	6	I, L	100–120	1250	5.51 (120) × 5.98 (152)	447 (7.3)	855 (389)	
NEC	NA	4	I, A, TS	70	1500	4.48 (114) × 3.98 (101)	251 (4.1)	290 (132)	

R—radial; I—inline; V—V-type; L—liquid-cooled; A—air-cooled; TS—two stroke; O—horizontal opposed; F—fan-type; Ro—rotary; NA—not available. Open figures—U.S. Customary; figures in parentheses—metric. Weights are dry unless otherwise specified.

THREE

The separate-cylinder liquid-cooled inline engine.

German Engines of World War I

At the outbreak of the First World War the world's most efficient and reliable aircraft engine was the 120-hp Austro-Daimler. This water-cooled inline six, the work of Dr. Ferdinand Porsche, was the prototype of all of the Central Power's wartime engines and of many built elsewhere (Fig. 3-1). Its crankcase was a shallow aluminum casting of roughly cylindrical shape. Through the use of dry-sump lubrication, in which a small pump (the *scavenger pump*) removes surplus oil from the crankcase to an external tank from which a pressure pump supplies the bearings, the deep crankcase of many other engines was avoided. Bolted to this crankcase were the cast-iron cylinders, each a separate and independent unit. The cylinder heads were not detachable, since they were made in unit with the cylinders. Water jackets made of sheet-steel stampings (beginning in 1914: previously the water jackets were electro-deposited copper) were welded to each cylinder. Overhead valves were set transversely to the centerline and were angled outward at about 30° from the vertical. They were actuated by pushrods from a camshaft in the crankcase. The pistons were iron, as was common at the time, and were very long in relation to their diameter. They drove the crankshaft through connecting rods that also were quite long and to modern eyes, spindly. The crankshaft itself was very heavy and relatively rigid. It was offset 0.709 in. (18 mm) from the centerline of the cylinders, in the interest of smoothness, and it ran in eight main bearings, the front one being a double-ball thrust bearing set ahead of the timing gears.

The whole design made sense in terms of the technology of the times, though some of its features might horrify a modern engineer. The lack of rigidity that resulted from the absence of a supporting cylinder block and from the shallowness of the

FIG. 3-1 120-hp Austro-Daimler, 1913. This, and the similar Mercedes, were the prototypes of the great majority of the German aircraft engines of the First World War.

crankcase did no harm. Because of the smoothness that resulted from the low compression ratio, the long stroke, and the long connecting rods, the sturdy crankshaft did not require a stiff structure to support it against breakage. The rocker arms and valve springs were exposed, but this meant that they were well cooled, and apparently the air over northern Europe was clean enough for these components not to need protection against dust. The overhead valve construction, while not necessary for maximum power, was congruent with one-piece separate cylinders. The separate cylinders eased maintenance problems because trouble in one cylinder did not necessitate tearing down the whole engine.

Its operating economy was excellent, with a specific consumption of 0.54 lb (0.25 kg)—better than a good many of today's engines. Good fuel economy results from good combustion-chamber shape, high compression, low mechanical and fluid friction, effective ignition, and proper operating temperature. Except for high compression—and even this, at nearly 5 to 1, was high for the period—the Austro-Daimler and many of the engines derived from it rated high in all these areas.

With a piston displacement of 850 cu in. (13.9 L), it was a decidedly big engine for its power output. It had to be big to produce 120 hp at its low speed of only 1200 rpm. This low speed, with its concomitantly high propeller efficiency, plus its low fuel consumption, went far to compensate for

its 575-lb (262-kg) weight. Because less weight in fuel had to be carried for a flight of a given duration, the Germans were able to obtain fairly good aircraft performance despite their heavy engines.

British Developments

This untidy but accessible layout was bound to appeal to designers in a nation that liked to put water pipes on the outside of houses so that they would be easier to get at when they froze, and the British did in fact have a long love affair with this type of construction. But so did many others. In France, Italy, the United States, and elsewhere, major engines for some time to come were developments from the basic Austro-Daimler pattern.

The Scottish Beardmore was a very close copy of the Austro-Daimler, even to the point of having the same 5.12 × 6.89 in. (130 × 175 mm) bore and stroke, but it somehow managed to give 135 hp at the same 1200 rpm. It apparently benefited from the peculiar British genius for getting more out of a copied design than its originators could. The next Beardmore had a larger bore and used dual carburetors instead of the original single water-heated one, and it gave 160 hp. Both versions saw service in the F.E. 2b pusher fighter and in the Armstrong-Whitworth F.K. 8 and gave generally satisfactory performance. The Beardmore later became the basis for even larger sixes; but since these engines used cast cylinder blocks, their story belongs to Chap. 5.

Rolls-Royce brought the separate-cylinder engine to its wartime peak of development in Britain. It began modestly by building[1] the Hawk, a 302-cu-in. (5-L) six, of 75 hp at first but soon raised to 100. Not powerful enough for combat use, it found useful employment in the hideous but effective D.H. 6 trainer and in early blimps (Fig. 3-2).

Having served its apprenticeship with the Hawk, Rolls next built a fine small twelve, the Falcon. This was a 280-hp engine using the same bore as the Hawk with a shorter stroke. In going to 12 cylinders, Rolls wisely demonstrated a distaste for the monster cylinders that others were employing in their high-powered sixes. I was told many years ago by a General Motors engineer that the "desire for simplicity isn't always a sign of good engineering. It's more often a sign of distrust in

[1]The Eagle was *designed* first, then the Hawk, and the Falcon last; but they went into production in the order given.

your manufacturing capabilities." It is notable that engine builders who had confidence in their ability to make reliable components had never feared the complications of multicylinder engines. Rolls and Pratt and Whitney exemplified this. Both preferred to use more cylinders rather than larger ones, and both had fine reputations for reliability.

At 2250 rpm, geared down to 1327 at the propeller, the Falcon was 1916's highest-revving engine. At first it lacked reliability. Aviators who had been flying ahead of Beardmores in their F.E. 2b's did not appreciate the Falcon's extra power when it had to be paid for in reduced dependability. However, in what became typical Rolls-Royce fashion, factory technicians were sent into the field, and the Falcon's troubles were soon tracked down and remedied. Used in such planes as the immortal F2B Bristol Fighter, the Falcon became one of the war's best engines (Fig. 3-3).

With the Falcon straightened out, the Rolls people then put their experience into the larger Eagle. This 360-hp twelve made its contribution to Rolls' reputation for excellence—and for disinterest

FIG. 3-2 Rolls-Royce Hawk, 100 hp, about 1916. The rosette-shaped plates just below the valve springs are exhaust port covers, used for storage only. The exhaust manifold would go on in their place. Since Britain had better use for Rolls' skills than for training plane engines, only 200 of these were built. *(Courtesy of Rolls-Royce Limited.)*

FIG. 3-3   Rolls-Royce Falcon, 265/280 hp, 1917. Poetry has been written about the joy of flying a Bristol Fighter with Falcon power. The objects at the front are carburetors. Rolls' carburetors had barrel valves instead of butterfly throttles. *(Courtesy of Rolls-Royce Limited.)*

in neatness. Everything possible was hung out in the open where it could be gotten at, at least once the superposed shrubbery had been cleared away (Fig. 3-4). It had a stiff cast-aluminum crankcase of what Rolls called its "cellular" design and was geared down from a 2000-rpm crankshaft speed to 1080 at the propeller. The reduction gearing was of the epicyclic type, something that by no means everyone has been able to make work well (Fig. 3-5). The problem of getting even fuel distribution among 12 cylinders was solved by using no less than four carburetors, but, as they were Rolls-Royce carburetors, they never gave a bit of trouble. Valve operation was by an overhead camshaft for each bank of cylinders, the overhead cam being a change that many builders made to the basic Austro-Daimler concept. The main point of the overhead-cam design may have been the simpler crankcase casting that resulted from not having to accommodate a camshaft in the case. It is worth remembering that aluminum was only about 20 years old as a commercial metal in 1917. Much of what was known about good foundry practice in iron could be applied to aluminum, but by no means all; the lighter metal had its own characteristics which had to be learned.

In early V-type engines the connecting rod setup was different from what we are accustomed to today. Modern practice is to stagger each bank of cylinders a little in relation to each other, allowing the connecting rod bearings of each pair of cylinders to be side by side on a single crankpin. A better but more expensive way is to use "fork-and-blade" connecting rods, with the rods from one side straddling those from the other. Both these methods call for bushings whose diameter is large in relation to their length, but before 1920 engineers did not know how to design such short bearings for good oil retention. Largely because of this, the favored solution was to use link rods for one bank of cylinders, somewhat as is done on radials (Fig. 3-6). This arrangement made one set of rods travel in a path that was not quite circular and led to one bank of cylinders having a different vibration pattern from the other. As vibration-absorbing engine mounts were not developed until several years after the First World War, it may be thought that early aircraft must have vibrated excessively, but this was not the case. The explosions were relatively soft, the engines themselves were heavy enough to suppress much of the vibration, and the wood-and-wire structures were better dampers of vibration than were the metal airframes of later times.

Since a 360-hp engine is much too large to start by swinging the prop by hand, Rolls included an electric starter. Even so, the weight, at 900 lb (409 kg), was not excessive. The resulting weight-to-power ratio of 2.5 lb/hp (1.14 kg/hp) was close enough to that of the later U.S. Liberty to make one wonder why we in the United States bothered to design that engine instead of building Eagles under license. The 360-hp Rolls delivered as much thrust as the 400-hp Liberty did because of being a geared engine with a slower-turning propeller, and the planes that used both engines performed at least as well with the Eagle as with the more powerful Liberty. The fact is that we might have built Eagles instead of Liberties had the Eagle been easier to build. Unfortunately, to make good Eagles called for workers and supervision of Rolls-Royce quality, and these just weren't available in adequate quantity. Rolls knew this. Its chairman said that he'd go to prison before he allowed anyone else to build Rolls engines, war or no war.[2] In view of what happened when subcontractors built Hispano-Suizas (see Chap. 5), he probably had the right idea.

The best-known Eagle applications were in the D.H. 4, the light bomber that could outrun Fokkers, and in the Vickers-Vimy heavy bomber. In some applications one weakness, not really Rolls' fault, developed: many cooling systems proved to be somewhat leak-prone. In accordance with Finnagle's law,[3] leaks could be counted on to occur at the worst possible time and place. Hawker and Greve, trying to fly the Atlantic in an Eagle-powered single-engine Sopwith, found this out; fortunately for them, the Sopwith floated well. All three British Atlantic crossing attempts in 1919 used Eagles, and two of them got Alcock and Brown across from

[2]He may have been referring to combat engines only. All Hawks were produced by Brazil, Straker and Co., the firm that later became Cosmos Engineering and developed the Jupiter radial.

[3]Often miscalled "Murphy's law" in recent years. Murphy's law is actually a mere corollary of Finnagle's, stating that "if a part can be assembled backward, someone will assemble it backward." See, for example, the Murphy's Law Room at Hamilton-Standard, where engineers check out new designs by trying to assemble them backward.

FIG. 3-4   Rolls-Royce Eagle VIII. Carburetors all over the place, ignition from four 6-cylinder magnetos, but it was as reliable as many and more so than most. *(Courtesy of Rolls-Royce Limited.)*

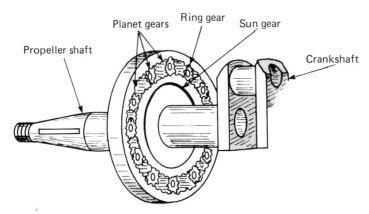

FIG. 3-5   Straight-cut epicyclic reduction gear. The crankshaft turns the sun gear, the ring gear is fixed to the inside of the crankcase, and the carrier for the planet gears is part of the propeller shaft. In the early Rolls-Royce version, the crankshaft drove the ring gear, the sun gear was fixed, and there were only three planet gears (or pinions). The above arrangement was more common—and far easier to draw!

Newfoundland to Ireland. Although an arduous in-flight repair was required, the fact that it did power the first nonstop transatlantic flight must stand forever to the credit of the Eagle and the Rolls-Royce firm (Fig. 3-5).

Rolls followed the Eagle with a monster, appropriately named the Condor. Again a twelve, it developed 600 hp—later raised to 670—from 2138 cu in. (35 L). It saw no wartime service and little post-war use. Its applications were confined to two types of aircraft that needed the largest available engines: the carrier-based torpedo plane and the large single-engine passenger transport. The passenger planes were more widely used in Europe than in America. In the United States we tended to favor smaller trimotors for routes that did not justify a normal multiengine craft (though a number of really tiny trimotors were built in Europe).

Although passengers and regulating agencies are reassured by multiengine planes, it is by no means certain that such craft are always safer. A contrary argument put forth by Canadian bush pilots is interesting and quite possibly valid. They say that, statistically, an engine will fail once every so many hours of flight, and so having two engines doubles the chance of an engine failure on any given flight. If in-flight failures were the major problem, a twin would still be safer because flight can be maintained with one engine out; but most engine failures occur on takeoff, where all available power is needed. In the United States a government ruling rendered the question moot. After several crashes of Lockheed Orions, two or more engines were made mandatory for all scheduled passenger service. This had the interesting side effect of making some big Fokker Super Universals available cheap for rum-running during the last years of Prohibition.

A Condor figured in an odd happening. In May 1927, a Condor-powered Hawker Horsley torpedo bomber left England and headed east in an attempt on the world's distance record. Some 34½ hr later it was forced down in the Persian Gulf after flying 3350 mi (about 5400 km). This was indeed a new distance record, but within just a few hours, the new record was broken by Lindbergh's arrival at Paris!

The last great British wartime engine was the Napier Lion. The first version used cast-aluminum cylinder blocks, but these were not satisfactory, and the Lion reverted to separate-cylinder construction for the rest of its long production life. The Lion was well named. It was big, strong, beautifully engineered, and capable of amazing output. The first Lions developed 450 hp, and the last, in a special racing version, no less than 1320! It was the outstanding large engine of its time in all ways but one—its inherently excessive frontal area. The Lion was a W-type, called a "broad arrow" in Britain; that is, it had three banks of cylinders, one vertical and the others on each side at 45°. Numerous photos of its innards have been published, and they show the first aircraft engine whose moving parts look reassuringly strong, yet it was far from being excessively heavy. Externally it was, as usual with British engines, rather baroque at first, but eventually it was considerably cleaned up. After an evolutionary period of change and modification, the definitive version combined the separate cylinders of current practice with an aluminum head spanning all 4 cylinders of each bank. This construction avoided the foundry problems that a full cast block would have invited while facilitating the use of a covered and protected valve gear.

Just too late for wartime service, the Lion competed very successfully with the Rolls products in large single-engine planes, was widely used in multiengine flying boats, and, surprisingly for what was basically a big-plane engine, won the Schneider Cup in a Supermarine S.5 in 1927. A Lion even powered a land speed record car, Sir John Cobb's Napier-Railton. It was a great engine (Fig. 3-7).

German Engines, 1914-1936

The Germans adhered closely to the 6-cylinder formula. There was a Benz V-8 that has been described as an unsuccessful copy of the Hispano-Suiza; unsuccessful it was, but its construction followed the separate-cylinder pattern, not the Hisso's cast-block design. Mercedes built an equally unsuccessful straight eight, heavy and inefficient like almost every other unsupercharged straight eight ever built; and there were the Oberursel and Siemens rotaries. Except for this very short list, every German engine, whether for war planes, trainers, or Zeppelins, was a six.

In addition, every German engine used separate-cylinder construction. The cast-aluminum blocks used elsewhere and the stationary air-cooled engines built in Britain and France were never even seriously experimented with. There were good reasons for this seeming lack of progressiveness. German technical excellence has often been conservative in nature, a matter of getting the most from an accepted concept rather than of pursuing new ones. German supplies of many materials, including aluminum, were very restricted during the war. The armed forces' demands for human resources meant that industrial production had to be based on

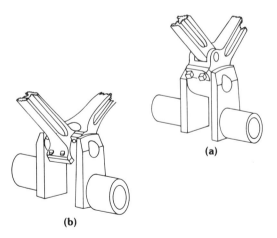

FIG. 3-6 Connecting rod arrangements for V-type engines. (*a*) Master-and-link type. Plenty of bearing area on the crank, but a little uneven in operation. (*b*) Fork-and-blade type. Theoretically perfect but expensive to make. Oil retention may be a problem for the narrow bushings of the fork rod.

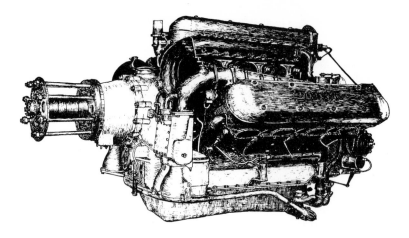

FIG. 3-7   Napier Lion, 12 cylinders, 450 hp. A magnificent engine but bulky. *(Courtesy of the author.)*

exploiting the skills available from a limited work force, and design changes that called for retraining were unacceptable. Finally, German tactical doctrine for fighter planes was largely defensive and did not demand engines of the highest power-to-weight ratios. On the whole, the inline sixes served the Central Powers well.

Most famous of these was the 180-hp Mercedes, a 1917 design that served as the engine of choice for German fighter planes during the last 1½ years of the war. It weighed about 700 lb (318 kg) and delivered its power at 1500 rpm from a displacement of 894 cu in. (14.65 L). It was the principal powerplant for the Fokker D-VII and the late Albatroses and Pfalzes, and a comparison with the Hisso that was used in their best-known opponents is of interest. The Hispano delivered its 220 hp at 1600 rpm from 718 cu in. (11.77 L) and weighed nearly 200 lb (91 kg) less. It had a longer TBO, and its only point of inferiority was its greater frontal area. Except for the Fokker, German fighters had better-streamlined noses[4] than did Allied ones. The inline six encouraged this, whereas it is likely that designers confronted with the bulk of the Hisso V-8 didn't think it worthwhile to attempt anything clever in the way of cowling design and radiator placement.

[4]The Austrian ones did not. Like the Fokker, Spad, and S.E. 5, they had car-type radiators out in front.

While the Mercedes was similar in concept to the Austro-Daimler, there were numerous detail differences. Cylinders were steel instead of iron, and valve operation was by an overhead camshaft rather than by pushrods. Two valves were used per cylinder. To facilitate starting, the camshaft was axially shiftable to a "half compression" position. Compression ratio, at 5.3 to 1, was somewhat higher than in most Allied engines. This was workable because the benzol (synthesized from coal) that was blended with the German gasoline raised the octane rating slightly over all other gas except that made from Romanian crudes. The crankcase had a sloping bottom, and a small amount of oil stayed in a compartment at the back of the sump, but the Mercedes kept the scavenging pump and separate oil tank of the Austro-Daimler (Fig. 3-8).

The rest of the German engines were much the same. The Benz that interchanged with the Mercedes in D-VIIs had two camshafts in the crankcase, much like the old T-head car and boat engines, operating the valves through pushrods. Benz used cast-iron cylinders, apparently without incurring any weight penalty; its largest engine developed 230 hp and weighed 848 lb (385 kg) dry. Maybach used pushrods and specialized in engines intended to run for hours at a time in the futile Zeppelins that absorbed so much of Germany's scarce aluminum. When engine trouble developed on long Zeppelin raids, it was common for a mechanic to remove the offending cylinder and keep the engine running on the remaining 5! Austro-Daimler enlarged its basic design to 200 hp by 1918. Another and more Mercedes-like Austrian engine was the Hiero, used to power the effective, if strange, Phönix fighters. Basse und Selve, best known as a radiator maker, built a Mercedes-like engine with stirrup-shaped cam followers that could pull the valves shut if a valve spring broke. BMW, in Munich, produced engines that were somewhat neater than most and specialized in high-altitude models in the later part of the war.

Until the coming of the supercharger, all engines lost considerable power in the thin air of high altitudes. Mercedes, Maybach, and BMW built

engines that offered a partial solution to this problem through the use of oversized cylinders and above-normal compression ratios. Such engines would destroy themselves if operated at full throttle at low altitudes, and so they required careful piloting technique. Aircraft using them, of which the Rumpler C-IV with the 260-hp Mercedes D-IVa was the best known, had to take off and climb to about 6600 ft (2000 m) at part throttle. Above this height the throttle could be opened until, at altitudes close to the aircraft's ceiling, full power could be used. Thus powered and carrying oxygen for the crew, Rumplers could fly photoreconnaissance missions high enough to be completely safe from interception. The reluctance to use engines calling

FIG. 3-8 180-hp Mercedes six, exhaust side. As the Albatros photo indicates, the upper part of the cylinders was always left uncowled regardless of the attention devoted to other aspects of streamlining. The Austrians sometimes cowled them, but Austrian planes sometimes suffered overheating while German ones did not. Note the lever for shifting the camshaft. (*Courtesy of NASM-Smithsonian.*)

for careful handling in fighters contrasts with World War II ideas. The operating manual for an engine like the Pratt and Whitney Twin Wasp with two-speed, two-stage supercharger is a formidable document, but pilots were expected to follow it closely even in the heat of combat. The difference, of course, lies in the technical background of young men in 1940 as compared with those of 1917.

Defeat and the Treaty of Versailles changed conditions in Germany almost as much as the Germans claimed it did. As in the Allied countries, some firms, notably Basse und Selve, Argus, and Mercedes, dropped out of the aircraft engine game. Maybach built automobiles and a few Zeppelin engines, and Benz merged with Mercedes to form the present Daimler-Benz. Nevertheless, there was enough business to attract one new manufacturer, Junkers, to the field. BMW detuned its wartime 220-hp engine and renamed it the BMW IIIa and found a fair market among the new airlines that started up shortly after the war. This and its 150-hp successor, the BMW VI, remained in production until 1930. When manufactured under peacetime conditions, both demonstrated excellent reliability. One of them powered the Junkers "Bremen" in which the first east-to-west transatlantic crossing was made.

A peculiar problem emerged when some of these BMW-powered Junkers were imported to the United States, under the name of Junkers-Larsen, in the early twenties. In Germany, the use of synthetic fuel, and perhaps of an early form of synthetic rubber, allowed the use of rubber tubing for fuel lines. Those sent here, though, employed the then-standard copper tubes. The vibration of the big six in a metal airframe caused fatigue failures (then sometimes called "crystallization") of the copper, and in-flight fires resulted. There had been some thought that the metal Junkers would be less fire-prone than the D.H. 4's then in airmail service, and so these fires were a considerable setback to the cause of metal construction in America.

The most widely used German liquid-cooled engines of the twenties were probably the Junkers. While Junkers had not built aircraft engines before 1921, it had considerable experience in marine and other Diesel engines. Professor Junkers is, in fact, credited with the original idea of the "overcom-

pressed" engine for high-altitude work. The first important Junkers was the L-5, a big (280/310 hp) six that differed from the classic German formula only in using ball main bearings and in having only two valves per cylinder. These valves must have been extremely large. All previous engines had gone to four valves per cylinder once the output went above 30 hp per cylinder. A later version, the L-8, with four valves per cylinder and the very high output of 440 hp, was also built. Each of these engines had a 12-cylinder derivative, these being the L-55 of 500 to 600 hp and the L-88 of 700 to 850 hp. While these engines were far from being advanced technically, the most successful radically advanced engines of the period were also Junkers products; Junkers built the only Diesel engines ever to see widespread service. These are described in detail in Chap. 11.

Welded-Cylinder Engines in France

Wartime engine development in France was rationally coordinated with aircraft requirements. Trainers generally used air-cooled stationary engines. Pursuits used rotaries and Hispano-Suizas. Light bombers, observation planes, and the smaller twins largely employed the Salmson Canton-Unne water-cooled radial. Only for applications calling for maximum power were water-cooled inlines used. Chief among these was the 300-hp Renault, joined late in the war by the Lorraine. Just after the war, the Farman line became important.

The first water-cooled Renault was a V-8, but most production Renaults were twelves. They were not substantially different from the German types and gave the same good service. In speaking of an excellent European engine of those days, what is meant is more a matter of workmanship and less a matter of design than would be the case in America, or later in Europe. The machinery, the tooling, and the inspection systems were not the dominating factor in manufacturing quality that we expect them to be today. Quality was much more up to the individual workers and their immediate supervisors. As late as 1929, Brownback was trying to make a virtue of necessity by referring to the Anzanis he was importing as "these fine hand-built French engines." The best of them were fine indeed, but when a crate was opened, one could not be entirely sure whether the engine inside would be a good engine or a bad one. In those days, the consistency of American engines was a source of wonder to European mechanics.

The Renault twelve seemed to have an affinity for aircraft that enjoyed long production lives. No doubt the excellence of the engine contributed to the satisfaction that those aircraft provided. One of the first to use it was the Levy two-seat flying boat, a patrol craft that was produced without significant change from 1917 to 1922. Among landplanes, the Breguet 14 was the classic Renault user. Breguet had previously used the 220-hp RE 8 Gd V-8 in its BM V pusher light bomber and had experienced problems with fires on takeoff. It was fortunate that this did not put Breguet off Renault engines, because the Breguet 14 was one of the war's most successful two-seaters. A slightly back-staggered biplane of metal construction, strong and easy to fly, it stayed in production from 1917 to 1926 and in service until 1930. Over 8000 were built, and they were used by the Americans, the Poles, the Romanians, the Finns, and the Spanish as well as the French.

For the 14's successor, the equally renowned

19 of 1921, some Renaults were used, but most 19s received Lorraines of 450 to 500 hp. Lorraine is credited with having built its first aircraft engines in 1915 but did not become an important producer until the war was practically over. A look at a map suggests the reason: the old province of Lorraine was front-line territory until the Germans were pushed back. Apparently only its 275-hp V-8 saw wartime service, being used in the Farman F-50 twin-engine bomber of 1918.

The classic Lorraine (Fig. 3-9) was the type 12E. It was a W-12, corresponding in many ways to the Napier Lion though much less impressive looking. Homely or not, it was good enough to power Breguet 19s to some remarkable distance records. One of these was the first nonstop Paris–New York flight, achieved in 1930 by Costes and Bellonte. As the Breguet was a biplane of considerably lower aerodynamic efficiency than were the Bellancas and Lockheeds favored by Americans for distance flights at the time, this and other long flights in Breguet 19s speak well for the qualities of the Lorraine engine.

The 12E was perhaps not a separate-cylinder engine in the strictest sense, since its cylinders were

FIG. 3-9 Lorraine 12E, 450 hp, of the mid-twenties. One of these engines powered the first nonstop Paris-New York flight. A far better engine than it looked! *(Courtesy of NASM-Smithsonian.)*

built in pairs. Nevertheless, it fits more nearly into this category than into the *en bloc* one because its cylinders were (paired) weldments rather than sleeves in an aluminum block. Lorraine took this type of construction to its ultimate with its 650-hp 18-cylinder engine of the late twenties. This was another monster, displacing 2255 cu in. (37 L). It was little used, and by 1929 the firm moved to cast-block design with the Courlis, one of the very few aluminum-block W-type engines ever built.

Late Continental Separate-Cylinder Engines: Italian Developments

Meanwhile Renault continued with the welded-cylinder scheme, making V-12s of as much as 750 to 800 hp and cataloging them as late as 1936 before bowing to the inevitable. By the late twenties, these engines, like most separate-cylinder types, used cast-aluminum covers for the camshaft and rocker arms instead of having exposed valve gear. This same construction was used for the Mercedes F-2, a large V-12 that marked Mercedes' abortive reentry into the large-engine field. (See Chap. 10 for Mercedes' small engine.) An incongruous feature of this 800/1000-hp engine was that it combined the old-fashioned welded-cylinder technique with the use of magnesium, the most modern of metals, for the crankcase. Argus also made a short-lived comeback in the same year with its 700-hp A.S. VI. This was a V-12 that, like the Lion, used separate steel cylinders tied together with a single cast-aluminum head for each bank of cylinders. None of these engines was widely used.

It was in Italy that the hybrid steel-cylinder-with-aluminum-head construction reached its highest development and widest use. The Italian engines were largely the work of two firms that were both better known, at least outside their homeland, as automobile builders, Fiat and Isotta-Fraschini.

FIG. 3-10(*a*) Fiat A-12bis, a late version of the A-12, 300 hp. This engine had a mixed record in service; it was a disaster in France but served well in Caproni bombers. A six of this size was probably a somewhat marginal proposition; Hispano-Suiza went to an eight and Rolls-Royce to a twelve when this much power was wanted. *(Courtesy of Centro Storico Fiat.)*

Earlier, during the First World War, another firm, Ansaldo, had been at least as widely respected for the qualities of its aircraft engines as Fiat was, and more so than Isotta was. Ansaldo was a large shipbuilding firm that had moved into automobiles and thence into aircraft engines. Its best-known product was an excellent 260-hp six, very German in concept. If the figures are to be believed, the S.V.A. biplane that this engine powered was, at 143 mph (230 km/hr), the fastest of all Allied pursuit planes. The S.V.A., built by a subsidiary of Ansaldo, introduced the system of wing bracing by a Warren truss of struts (instead of wires) that continued in use for Italian fighters right to the end of the biplane era. After the war, Ansaldo faded out of the aircraft and automotive fields to concentrate on shipbuilding.

Fiat, like Rolls, began modestly with a 100-hp engine, the A-10. Its first significant wartime engine was the 6-cylinder A-12 of 225/268 hp. It was neater than most separate-cylinder motors and had an intake manifold whose shape showed its designers' concern with good fuel distribution. It had two odd features. One was that the valve gear was lubricated neither from the main lubrication system nor by oil cups, but by packing its casing with heavy oil every 12 hr of operation. The other oddity was its use of Dixie magnetos. These were American products of little repute at home; in fact, the Dixie magneto was one of the curses of the dreaded OX-5.

Fiat sixes were used in Italy's only other homegrown fighter, the Fiat Ballila. Some of the 300-hp A-12bis versions were exported to France, where they were the key element in an incident that I would like to know more about. Used in the Voisin XI pusher light bomber, the A-12bis developed a habit of catching fire in flight. After several deaths had resulted, some crews refused to fly them and were arrested. The Fiat-powered version of the Voisin was then taken out of service, but one wonders what happened to the fliers whose actions led to its belated condemnation (Fig.3-10).

At the end of the war, the A-12 was followed by the largest twelve ever produced in quantity, the 3488-cu.-in. (57-L) A-14. It had a stroke of no less than 8.27 in. (210 mm), making for an excessively high piston speed and for short bore life. Everything about the A-14 gives the impression of first-class

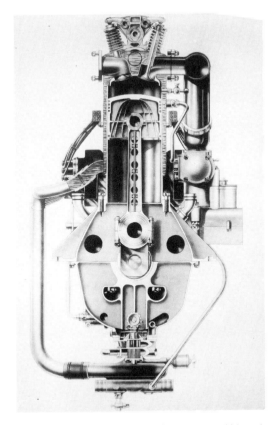

FIG. 3-10(b) Cross section of Fiat A-12. Although the detail engineering is more impressive than that of most such engines—notice the ribbing inside the piston and the small tube to bring oil under pressure from the crankpins to the wrist pins—this is typical of the way the separate-cylinder engines were constructed. (Courtesy of Centro Storico Fiat.)

detail engineering, but this was just too big an engine. Even so, some big Capronis used it during the first 4 years after the war.

Having gone to about the limit for a fully separate cylinder, Fiat next designed an entirely new type. In its A-15 (Fig. 3-11), cylinder blocks rather than separate cylinders were used, but they were not cast in aluminum. Instead, they were fabricated from sheet steel with a single water jacket enclosing the whole bank. This was an extraordinarily good looking engine. It had a clean gear case for its herringbone reduction gears (Fig. 3-12), a camshaft cover that continued the lines of the

FIG. 3-11   Fiat A-15, 430 hp. If this engine had lived up to the promise of its appearance it would have been a world-beater. But what sort of strange internal exhaust manifold is feeding from 6 cylinders to four ports? And how is the mixture brought from a carburetor, mounted low on the outside of the block, to three intake valves high on the inside? And what kind of odd drive actuates magnetos at the sides of the crankcase instead of at the rear, where one would expect to see them? The neat appearance seems to have been achieved at the expense of some internal funny business. *(Courtesy of Centro Storico Fiat.)*

cylinder blocks, and manifolding encased within the water jackets. It was in all ways the opposite of British practice and might in fact have profited from a dose of British patient persistence in debugging. Despite impressive specifications—430 hp from 1240 cu in. (20 L) at 2415 rpm geared to 1600 at the propeller—it had a short production life. Possibly, as in the case of the contemporary Curtiss D-12 of similar size and power, the reduction gears gave trouble. Despite the quietness of herringbone gears, a great many designers until very recently have preferred to accept the roughness of straight-cut gears as the price of greater strength.

In 1925 Fiat dropped its existing line and came out with an entirely new range of engines. All were twelves of classic separate-cylinder construction with an enclosed valve gear. They were less elegant than the A-15 but must have been better engines, because they enjoyed a long production life. Fiat cataloged them as late as 1936, although few were made after 1933. The new range extended from 430/450 hp for the A-20 through 700/750 hp for the A-22 to the 950/1000-hp A-25. In addition, there was the A.S. 5 racing engine, developed for the Schneider Trophy contest, which gave 1000 hp from only 782 lb (355 kg). Since the A.S. 5 had the remarkable (for 1929) compression ratio of 8 to 1, it must have run on "dope" racing fuel. This would have included, at that time, benzol, extra tetraethyl lead, methanol (to keep temperatures down), and ether (to promote volatility). At an earlier date there would have been some picric acid, a touchy explosive; later nitroglycerin might be added. When racers exhaust the possibilities of hopping up a given engine, they turn to trying wilder and wilder brews if the rules permit. Sometimes this helps; sometimes it just blows up engines.

Fiat's final racing engine was a tour de force of the first magnitude. Called the A.S. 6, this 24-cylinder monster was really two 12-cylinder engines built as one unit. Both were geared. Power was taken from the middle to two shafts rotating in opposite directions and mounted coaxially to drive counterrotating propellers. Counterrotation was badly needed when 2800 hp drove a 5500-lb (2500-kg) aircraft, especially in the days before the controllable-pitch propeller. This engine, after 2 years of frustrating effort, enabled Lt. Francesco Agello to set a record of 440 mph (710 km/hr) in 1934, in a Macchi-Castoldi seaplane. This record still stands for seaplanes and, in view of the time and money expended to achieve it, will probably continue to stand for all time (Fig. 3-13).

The Fiats and the Isotta-Fraschini Asso line were the mainstay of Italian military aviation during the late twenties and early thirties. The big A-25—at 3321 cu in. (54.42 L) it was close to the size of the earlier A-14—was especially noteworthy. As airframe construction had advanced since the days of the A-14, such large engines were now becoming practical. Both in Italy and elsewhere, the monster motors of the late twenties led to some very large single-engine planes. Italy's Fiat B.R. 2 and B.R. 3 bombers were among the most successful of these. Biplanes of truss-braced wing construction, they spanned 56 ft 9 in. (17.29 m) and weighed almost 10,000 lb (4545 kg). They were built in considerable numbers and saw many years' service with the Regia Aeronautica, the Italian Air Force. Caproni went even further in 1932 by converting its Ca 101 trimotor into a big single, the Ca 111. This used a 950-hp Isotta and weighed 11,795 lb (5361 kg). The biggest of them all came from Dearborn. In 1931 Ford substituted a big 650-hp Hisso for the three Wasps of its 5-D trimotor to make what it called the Ford Freighter. It spanned 77 ft 10 in. (32.77 m), grossed 11,000 lb (5000 kg), and was most elegant in appearance, but failed to sell. Airfreight in 1931? Surely Henry Ford, of all people, wouldn't build an aerial rumrunner?

Isotta-Fraschini was basically a manufacturer of large luxury cars, a fair number of which were exported to the United States during the twenties. It built its first aircraft engine as far back as 1911. In that year, the Isotta V-1 airship engine was the first gasoline engine ever (large, slow stationary engines excepted) to run for 24 hr under full load. The impressiveness of this achievement is somewhat lessened by the fact that the V-1, weighing 682 lb (310 kg) for an output of only 90 hp, was distinctly a derated engine; nevertheless, no other aviation engine had done as well up to then.

Isotta continued to build heavy engines, with iron cylinders cast in pairs, through the V-5 model of 1915. Then, with the 275-hp V-6 of 1917, the firm took a long step toward the continuous aluminum head that became almost an Isotta trademark. In the case of the V-6, an inline six, one head extended over each of the three pairs of cylinders. It was a somewhat fragile engine, Isotta having gone to a four-bearing crankshaft in its newfound enthusiasm for weight saving. Its model designations at this time were confusing. The V-6 was in fact an inline six, as was its successor, the 300-hp V-8. The V-8 was then the strongest Isotta, having seven main bearings and a strong crankcase.

For the V-10 and the Asso (Ace) 200, a cast-aluminum head spanning all 6 cylinders was used. This construction characterized Isotta-Fraschini engines for the next 18 years—until 1938—and it worked very well during that period. It was next used on the firm's only V-12, the 1925 Asso 500, one of the few engines to get below the magic specific-consumption figure of 0.50 lb (0.227 kg). Asso numbers, incidentally, gave the nominal horsepower of the motor.

For further power increases, Isotta used more cylinders rather than larger ones. Its most famous product, the Asso 750 of W-18 configuration, gave an extremely dramatic demonstration in 1930 of just how reliable a water-cooled engine of complex construction could be. In that year, 12 twin-hulled Savoia-Marchetti flying boats made a round trip—

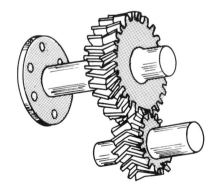

FIG. 3-12 **Herringbone reduction gears. For best strength, gears should have a tooth depth approximately equal to the face width. This is easy with straight-cut gears but impossible with the herringbone design.**

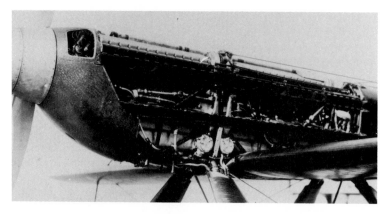

FIG. 3-13(a)   Fiat A.S.6 engine, 2800 hp, installed in the fastest seaplane ever—440 mph (710 km/hr). It took about 3 years of development effort to achieve this record. *(Courtesy of Centro Storico Fiat.)*

flying in formation, no less—from Italy to Rio de Janeiro under the leadership of General Italo Balbo. In 1933 they did it again, this time to Chicago and back. This performance was, in fact, too dramatic for Mussolini's liking, because Balbo was shortly shifted from his post of air minister to the relative obscurity of the governorship of Libya (Fig. 3-14).

Isotta was one of the first to build a 1000-hp engine, its Asso 1000. This was another 18-cylinder W-type, with a displacement of 3500 cu in. (57.38 L). This is even larger than the Fiat A-14 but, unlike that engine, was produced only in experimental quantities. It is a little hard to think of such giant powerplants as being the result of rational calculation of commercial possibilities. Perhaps they were

exercises in exploring the limits of the state-of-the-art, or they may have resulted from pressure originating with the military. American experience suggests the latter explanation.

Six Asso 1000s went into one of the worst-looking airplanes ever built, the Caproni Ca 90 P.B. This followed a pattern that Caproni had been using since it gave up on triplanes around 1923. A large wing sat on top of a boatlike fuselage, and, far above it, there was a smaller one. The interplane struts sloped sharply inward, and a nacelle carrying a push-pull pair of engines stood above the fuselage on struts. On each lower wing was another pair of Asso 1000s. The wingspan was 153 ft (46.63 m), and it weighed 66,000 lb (30,000 kg), making it

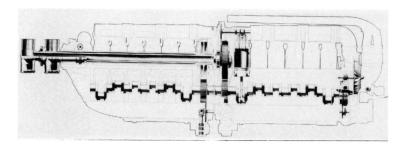

FIG. 3-13(b)   Cross section of Fiat A.S.6 engine. Complex in concept, but simple in execution—no herringbone gears here, just massive straight-cut gears. The supercharger takes the mixture from four carburetors set in line across the rear of the engine and is of surprisingly modest size compared to that on the Rolls-Royce "R." *(Courtesy of Centro Storico Fiat.)*

FIG. 3-14 Isotta-Fraschini "Asso 750," a strong contender for the title of "best separate-cylinder engine ever built," but nevertheless a good 5 years obsolete when built. The annular corrugations on the water jackets were to allow some flexibility in expansion, since the cylinders to which they were fastened got hotter than the jackets. All late welded-cylinder engines, including the PT-boat Packards, had this feature. *(Courtesy of NASM-Smithsonian.)*

the largest bomber in the world at the time. The whole apparition zipped along at 127 mph (205 km/hr).

There is little doubt that Italy led the world in large liquid-cooled engines between about 1928 and 1935.[5] Yet by the time of the Second World War, Italy had to go to Germany for the design of a liquid-cooled engine suitable for an effective fighter plane. At the outbreak of the war, other nations had V-12s of 1000 hp or more,[6] but Italy's best was the 835-hp Asso XI. R. The Italians never designed an aluminum-block engine until after

World War II. They were probably led astray by their success with fabricated-cylinder powerplants until it was too late to strike out in another direction. They took the opposite tack from that taken earlier by the British, who stopped production on the Eagle (at a time when that engine was about as good as anything in the world in its class) to start fresh with the Kestrel. The Kestrel led to the immortal Merlin, but the Asso 750, good as it was, led up a blind alley.

What eventually made obsolete the separate-cylinder engine was the superior strength and rigidity of the cast-block engine. The more power obtained from each cubic inch of displacement, the more highly stressed are all parts of the engine. Fatigue failure will be caused by flexing if rigidity is inadequate in a high-output motor, and so the output of a separate-cylinder engine is inherently

[5]These dates are based on the Kestrel being uncompetitive in power after 1928 and the Merlin coming out in 1936 (see Chap. 5).
[6]The French thought they had one, in the 12Y Hispano, but it proved unreliable at over 860 hp in fighters.

limited to a lower figure than that of a comparable cast-block engine. To illustrate this, two 1936 twelves can be compared. The Asso 750 R.C., by then a fully developed design, gave 950 hp for takeoff at 1700 rpm from 2865 cu in. (47 L). The Merlin, new that year, gave 1000 hp at something over 2000 rpm from 1650 cu in. (27 L). The supercharged Asso may well be taken as the epitome of good separate-cylinder design; it seems that if reliability were not to be compromised, specific output would peak out at around 1 hp for each 3 cu in. (49 cc) of displacement. In contrast to this, the Merlin had nearly twice the specific output in 1936 and, eventually, better than 1 hp/cu in. (16.4 cc) with complete reliability. While the Merlin's weight per cubic inch was higher than the Asso's, its weight per horsepower was lower, so the Merlin's strength was due more to the inherent advantages of cast-block construction than to heavy construction.

FIG. 3-15 Curtiss OX-5, 1915 to 1918—90 hp at 1400 rpm, about 400 lb (182 kg) with water, oil, and radiator. *(Courtesy of the author, taken at Glenn Curtiss Museum, Hammondsport, New York.)*

American Separate-Cylinder Engines

In the United States there were four important separate-cylinder engines. The first of these were the Curtiss OX-5 of 90 hp and its 100-hp companion, the OXX-6. The OX-5 was the culmination of an evolutionary process of Curtiss's V-8s. It was first built in 1910 and is still flying in antiques today. It had a wet-sump crankcase and steel cylinders with brazed-on Monel water jackets (Monel is a copper-nickel alloy). The overhead valves were actuated by pushrods and pull rods, the intake pushrods working inside the tubular exhaust pull rods. While the OX-5 was as reliable as most pre-war engines, by the time it reached the end of its long career it was probably the least reliable aviation engine in widespread use.

Its frequent failures came mainly from three things. The valve mechanism was neither durable nor reliable, the Dixie magneto was a poor thing, and the cooling system was leak-prone. The OX-5's reliability could be improved by substituting a Bosch or Splitdorf magneto and by using certain other proprietary replacement parts, but there was just so much that could be done. It remained a heavy, short-lived, single-ignition engine. In addi-

tion, the drag of the large amount of oil in the crankcase and the long, thin intake manifold made it a bear to start in cold weather. It was common practice to drain the oil after a flight and bring the oil indoors overnight to keep it from thickening (Fig. 3-15).

Although even then obsolescent, this engine was put into large-scale production in 1917 as the mainstay of the American wartime training program. The authorities had little choice but to do so. There was only one good domestic trainer in being, the Curtiss JN-4, and that plane had been designed around the OX-5. In the United States, the advances in engine design since the OX-5's introduction had not reached the horsepower range then considered appropriate for training planes. The motors used for training by the French and the British were either not that much better or, as in the case of the Rolls-Royce Hawk, were harder to build. Where our government must be faulted is in making too many. By the end of the war we had OX-5s coming out of our ears.

The 150-hp Wright-built Hisso was interchangeable with the OX-5 in the Jenny and "made an airplane out of it." The military therefore dropped the OX-5 as its basic training engine very shortly after the war. The result was that there were thousands of OX-5s, a high proportion of them brand-new, available cheap as war surplus. Until about 1928, therefore, OX-5s were the backbone of private aviation in the United States, and it is hard to say whether they were a blessing or a curse. They were extremely cheap, and many people could afford them, but, being so big and heavy, the OX-5 didn't lead to the design of good lightplanes. All the classic planes of the twenties used them: Waco, Curtiss Robin, Alexander Eaglerock, Brunner-Winkle Bird, and Fairchild/Krieder-Riesner, for instance. Every one of these planes was designed around the OX-5, an engine known to be obsolete even when the plane was on the drawing board. Most of them were good enough to be converted to the use of better engines and continued production with the better engines when the supply of OX-5s dried up. None of them were low-cost lightplanes in the sense that the British Gipsy Moth was—cheap and easy to operate. For the most part, pilots who owned OX-5-powered planes got "Limited Commercial" licenses and made their planes earn part of their keep by taking up joyriders on weekends.

On the face of it, there is something strange about the fact that all these planes performed as well as they did. They were all three-seaters, and the OX-5 weighed about as much as a modern 90-hp engine plus an average adult. Thus the gross weight of the aircraft corresponded with that of a four-place plane. If one asked a modern designer to design a 2000-lb (909-kg) airplane to be powered by a 90-hp engine, the designer would tell you that it couldn't be done. How did these aerodynamically unrefined planes (except for the Bird and the Robin, perhaps) (Fig. 3-16) fly as well as they did on OX-5 power?

The answer lies in the fact that the OX-5 was a 1400-rpm engine, while a modern 90-hp engine turns around 2450 rpm. Torque, not horsepower, turns the propeller and produces thrust. The OX-5 swung a big prop and produced much more thrust than a modern motor of the same horsepower. A 2450-rpm engine would have to develop about 113 hp to produce the same thrust as the slow-turning 90-hp pre-war design did. It may have had only 90 horses, but they were big horses.

Another thing that made OX-5 performance acceptable was that the JN-4D constituted a standard of performance that was easily exceeded. For example, the Bird, last and best (though not most rugged) of the OX-5-powered three-seaters, weighed fully loaded only a little more than a Jenny did empty, and its aerodynamic efficiency was much higher. Its makers claimed a top speed of

FIG. 3-16 Curtiss Robin, 40-ft (12-m) span, 90 hp, three place, 1928. Seated four when Challenger-powered. As the airplane that brought the comforts of what Wiley Post called "indoor flying" down to the OX-5 flier's price range, the Robin sold so well that Curtiss established a subsidiary, Curtiss-Robertson in St. Louis, to build it. *(Courtesy of NASM-Smithsonian.)*

105 mph (169 km/hr) and a climb of 720 ft/min (220 m/min), and these may have been honest figures (manufacturers' claims in those days usually were not). In comparison, a JN-4 topped at around 75 mph (121 km/hr), and its rate of climb was best described as imperceptible.

Something of a cult surrounds the OX-5 today. For instance, there is an "OX-5 Club," open to those who have flown behind one. However, it is something like the cult of the Model A Ford in that both were considered pretty poor affairs in their prime and were largely used by those who couldn't afford anything better.

Hall-Scott also built engines for trainers but, after going as high as 150 hp in its sixes and making an unsuccessful small twelve, lost the services of Mr. Hall to the Liberty program for the duration. Its chief wartime product, the 4-cylinder 100-hp L-4, was made in sufficient numbers for several hundred to be sold postwar as surplus. Few were used in aircraft, but they had some popularity as powerplants for racing cars and boats (Fig. 3-17). After the war Hall-Scott left the aircraft engine business and turned to the less glamorous but more rewarding truck and boat engine field, gaining some reputation for good rumrunner engines. The chief importance of the Hall-Scott line lies in the contri-

FIG. 3-17   The fate of many a war-surplus engine. A Hall-Scott L-4, partly converted for marine use (water-cooled manifold) reposes in a barn. *(Courtesy of the author, from the Fyfield Collection.)*

bution that its chief designer made to the Liberty as a result of his experience with these engines.

Packard also built water-cooled engines. Under the vigorous leadership of its chief engineer, Jesse C. Vincent, Packard brought out the world's first quantity-production 12-cylinder car in 1916. Evidently Vincent's belief in the twelve carried over to the aviation field, because in early 1917 the firm built a 250-hp twelve for aircraft. This engine was rather like the Rolls-Royce Falcon but was somewhat easier to build, and the Liberty was to a large extent a scaled-up version of it. Packard abandoned its own engine work, during the war, in favor of Liberty production, but as soon as peace came, it resumed its efforts toward becoming a major builder of engines of its own design.

These efforts met with more failure than success. Packard's first problem was to decide on the best horsepower class to aim for. America had more OX-5s than it knew what to do with; we had plenty of Wright-Hissos to meet our needs in the 150- to 300-hp range; and we had enough surplus Liberties for our requirements in the 400-hp class for some time to come. In addition, the Curtiss K-12, as it then was, appeared likely to become the Liberty's replacement if its problems could be solved. All these were water-cooled engines, and had Packard explored the air-cooled field, the results might have been different. However, Packard's expertise, such as it was, lay in the area of the separate-cylinder water-cooled engine.

Packard's strategy was to scale up and modernize the Liberty. The resulting engine was 42 percent larger than its parent, at 2500 cu in. (41 L), and delivered 700 hp in the original version. In later versions this rose to 750, then 800, and culminated in America's first 1000-hp engine. While these advances in power were to some extent technical triumphs, they were flawed by the unsatisfactory reliability of the A-2500 (later the V-2500).

The big Packards were raced during the early twenties when the Armed Forces were financing high-speed experimental planes, but they seldom won. For the most part they broke down and "DNF'd," but even when they completed a race it was usually behind D-12-powered planes. The only place where the Packards came close to making good was as the powerplant for torpedo planes.

The British had shown the way in dropping torpedos from aircraft but had only been able to do it by building small torpedos of about 1000 lb (455 kg). Our Navy was anxious to use more powerful torpedos and, thanks to a good design from Douglas, succeeded to a certain extent. The Liberty-powered Douglas DT was, however, slow and could not lift both a torpedo and enough fuel to give it a useful combat range. Clearly more power was needed, and the big Packard looked like the answer (Fig. 3-18). Boeing and Martin both built big folding-wing biplanes of satisfactory performance, powered by Packards of 710 to 770 hp but the engines were not as good as the aircraft. Experience with these Packards contributed heavily to the Navy's 1925 decision to standardize on air-cooled engines and to its support of Pratt and Whitney's development of the Hornet. Perhaps the most useful result of Packard's efforts was the impetus given to large radials.

After 1928 there were no Packard aircraft engines on the market, but the basic design did not pass completely from the scene. As insurance against the possibility that the then-new Allison might not live up to the hopes held for it, the government encouraged Packard to revive the design in supercharged form around 1938. Fortunately, the Allison turned out reasonably well, but Packard's effort was not wasted. The "new" 1000-hp Packard turned out to be just what was needed to power PT boats. I worked on a pair of these engines around 1950 and was struck by the visual evidences of Liberty ancestry. The main visible difference was the covered valve gear.

The Liberty was in all ways the most important American separate-cylinder engine. It is difficult to be objective about something that has been both praised and damned to the extent that the Liberty has. Briefly, the Liberty was a big, up-to-date, but conventional engine designed under government auspices to meet a wide range of wartime requirements. Both its good and its bad points derived from, and are consistent with, this set of facts.

The government's 1917 aircraft engine program had to encompass several facts. First, engines of three different sizes seemed to be required. Second, America possessed great automobile engine production capacity but very little aircraft

FIG. 3-18 Packard V-2500. The sign calling it a −3A is thought to be in error. Note the rather excessive size of the gear case employed—it would seem that it should not have been necessary to use such massive gears. (Courtesy of the author, taken at Bradley Air Museum, Windsor Locks, Connecticut.)

engine production capacity. Third, while the Hispano-Suiza was the world's best aircraft engine, automobile firms abroad had made a hash of trying to produce good Hissos. Fourth, the Hisso's design was not amenable to much enlargement. Fifth, the "Mercedes-type" cylinder construction was well within the capabilities of many automobile builders. Sixth, if a satisfactory Mercedes-type cylinder could be developed, it could be made the basis for a number of different engines by just varying the number of cylinders used. Seventh, between Packard and Hall-Scott we had available some recent experience in designing separate-cylinder engines. Given these parameters, the engine that resulted could only have been something close to what emerged as the Liberty.

Those who were called upon to implement these objectives were well chosen: Jesse G. Vincent of Packard and E. J. Hall of Hall-Scott, the two Americans with the most experience in the type of engine the government wanted. The story has often been told of how these two men settled in a Washington hotel suite and had drafting materials and draftsmen sent in, how they designed the engine between May 30 and midnight of June 4, and how the first engine was completed and running by July 3. In truth this was a remarkable achievement, hardly diminished by the fact that the actual design was an amalgam of what Packard and Hall-Scott had already done.

The engine that emerged had only two features that departed from accepted European practice. Instead of magnetos, it used coil ignition, like an automobile, and the angle between the cylinder banks was 45° instead of the normal 60°. Coil ignition was adopted out of well-founded distrust in our ability to build good magnetos in quality, and the narrow angle was adopted to reduce frontal area. Except for these things, the Liberty could have been designed in any of the combatant nations; yet it had, in its basic 12-cylinder form, the best weight-to-power ratio of any engine to date. It displaced 1649 cu in. (27 L) and developed 400 hp for a weight of 844 lb (384 kg), a weight-to-power ratio of 2.44 lb/hp (1.109 kg/hp). For the first time, the Manly engine had been surpassed in lightness for power.

On the whole, the engine worked well from the beginning, though many changes had to be made as experience accumulated. The bearing design proved to be marginal and had to be rescued by a new manufacturing technique developed by Ford, then at the height of his powers. Some builders' water jackets tended to develop

cracks, and the welding sequence had to be modified. The original "dipper" oiling system for the connecting rods was changed to a full-pressure oiling system. The reliability of the Liberty has been much debated, but the fact seems to have been that Liberty quality and reliability varied considerably from maker to maker. A good Liberty was extremely reliable, a bad one pitiful. Carefully selected Liberties flew nonstop across the continent (Fig. 3-19), made the first Atlantic crossing, stayed up 36 hr in the first in-flight refueling experiment, and flew Army Douglases around the world; but run-of-the-mill Liberties in airmail service were somewhat troublesome.

The worst Liberties came from Packard, a fact excused by Packard's being assigned to carry the main burden of development and modification; the best were built by Lincoln; and those from Ford fell in between these extremes. The Lincoln Liberties owed their excellence to the character and intellect of Henry Martyn Leland, one of the great men in the history of the automobile. Leland had founded Cadillac after some years of building motors for others. Under his leadership, Cadillac had

FIG. 3-19 The state of the art in 1917—the Liberty 12. *(Courtesy of the author.)*

been the only firm to win two of the coveted British Dewar Trophies, one for interchangeability of components and one for its first V-8. When he was in his seventies, Leland sold his share in Cadillac to start a new firm with which he believed he could make a greater contribution to the war effort than was possible through General Motors. His Lincoln Liberties were not only the best, they were also the cheapest. A grateful government responded to this by filing enormous back tax claims against Lincoln, an action which brought the independent existence of the firm to an end in 1923.

The original plan was to build the Liberty as a four, a six, an eight, and a twelve. The first Liberty completed was an eight. It was mounted on the back of a truck and carried to the top of Pikes Peak for altitude testing, a far cry from today's use of altitude test chambers and flying test-beds. It ran well but vibrated too much for use in an aircraft, and no more were made. The six was a good enough engine, but only 52 were made; Curtiss used a few. It had been bypassed by the fact that Wright had finally solved its problems and was producing good Hissos to meet our needs in the 150-hp range. While 2 fours were built and forgotten, no less than 20,478 twelves were made, 2543 of them after the Armistice (disgracefully, but Wilson's "liberal" administration was tenderly solicitous of business profits in practice and was slow to cancel many war contracts). The British bought 980 at $7300 each with spares, and the French got 405 at $6000 each.

The Liberty twelve was the only American-built engine to see combat service. It powered the British-designed, American-built D.H. 4s that were America's sole contribution in materiel to the Allied air forces. The D.H. 4 was built in large numbers and served for at least 10 years after the war in the Army Air Corps, as well as being the mainstay of the early airmail. There were better planes. The use of the Liberty instead of the original Rolls changed it from an exceptional aircraft into a mediocre one.

The number of Liberties used in D.H. 4s was greater than all other applications put together, but Liberties did find their way into numerous other planes. The British apparently used all theirs in multiengine planes, such as Vickers-Vimy bombers and various large flying boats. Four of them pow-

ered the NC-4 across the Atlantic, and two were used in the Martins that formed the Army's first real bombing force. A Liberty powered the Dutch-built Fokker T-2, now in the NASM (National Air and Space Museum), in which McReady and Kelly made the first nonstop transcontinental flight.

Liberties were expensive to operate. It took a lot of skilled maintenance labor-hours to keep them running. Engine troubles ended McReady and Kelly's first two tries, and only an in-flight electrical repair enabled the third try to succeed. It took spare engines and a costly traveling circus of repair facilities to make the Army's round-the-world flight a success. As a result, there were only three civilian applications of any significance. "Bill" Stout used one in the Ford Trimotor's predecessor, the Stout "Pullman," and Boeing and Douglas put them into pretty good mail planes.

Yet the Liberty was a long time dying. The performance that a Liberty installation gave to a fast "freight boat" went far to reduce the attrition caused by the Coast Guard and made rumrunning a paying proposition. Three of them powered the crude White Triplex Land Speed Record attempt car in which Ray Keech was killed. Oddest of all, the Liberty was put back into production in England in 1938 as an engine for tanks. The "Cruiser" and the "Cavalier" actually saw service in World War II with Liberty power. Over 7000 new Liberties were built by the Nuffield organization (MG, Morris, etc.) before production ended in 1942. In the United States there were attempts made to produce an updated design using Liberty components. Wright experimented with new cylinders and complete cast blocks to use the Liberty base, but the most successful modification was Allison's conversion to inverted operation (Fig. 3-20). The resulting high thrust line made Loening's "flying shoehorn" (the first production amphibian) practical, and a fair number were built.

The Slow Death of a Concept

As we will see, after about 1918 the Hispano-Suiza-originated concept of aluminum blocks and steel sleeves led to generally better engines than did the separate-cylinder concept. The Hisso and the Curtiss D-12 had demonstrated this quite conclusively

FIG. 3-20   The actual liberty engine that powered the first nonstop transcontinental flight. One of the few Liberties built by General Motors, it has been called a "Buick Model A." (*Courtesy of NASM-Smithsonian.***)**

by 1923. Thus the persistence in certain quarters of the separate-cylinder engine needs explaining. In all probability there were two principal reasons for this persistence. One was the NIH[7] syndrome. In Germany especially, the welded-cylinder engine was thought of as a homegrown type while the aluminum-block engine was considered a French notion. Possibly more to the point, making good castings of the size needed for engine blocks was not easy in aluminum. Some nations seemed to learn the art and mystery of casting the light metal better than others. The Germans were still putting

spongy aluminum castings into Mercedes-Benz cars in the late 1950s, for instance, and poor aluminum castings helped sink the Borgward firm at about the same time. The American record was spotty. We were able to cast good blocks for the D-12 in the early twenties, yet the American-built Cirrus suffered from porous cast-aluminum cylinder heads. It may be, then, that the continued use of separate welded-steel cylinders was not necessarily due to technical backwardness or to mere stubbornness. In large part it may have been because of a realistic assessment of available capabilities. In the end, of course, the aluminum-block scheme won out everywhere, but its total victory was a long time coming.

[7]NIH stands for "not invented here."

Data for Engines

Maker's name	Date	Cylinders	Configuration	Horsepower	Revolutions per minute	Bore and stroke, in. (mm)	Displacement, cu in. (L)	Weight, lb (kg)	Remarks
Austro-Daimler	13	6	I, L	120	1200	5.12 (130) × 6.89 (175)	850 (13.9)	575 (262)	
Beardmore AR-3	14	6	I, L	135	1200	5.12 (130) × 6.89 (175)	850 (13.9)	600 (273)	Weight has been given as 600 lb (272 kg), but this seems too low
Beardmore AR-3	16	6	I, L	160	1250	5.63 (142) × 6.89 (175)	1025 (16.7)		
Rolls-Royce Hawk	15	6	I, L	75	1350	4 (102) × 6 (152)	453 (7.4)	405 (184)	1917 hp: 100 at 1500 rpm; date is for design, not production
Rolls-Royce Falcon	16	12	V, L	205	1800	4 (102) × 5.75 (46)	867 (14.2)	NA	1918 hp: 285 at 2250 rpm; produced before Hawk Initial (1916) hp was 266
Rolls-Royce Eagle	15	12	V, L	360	2000	4.5 (114) × 6.5 (165)	1240 (20.3)	900 (409)	
Rolls-Royce Condor	18	12	V, L	650	1900	5.5 (140) × 7.5 (191)	2138 (35)	1300 (591)	Geared (usual) model used 16-ft (4.9-m)-diameter propeller
Napier Lion	18	12	W, L	450	1925	5.5 (140) × 5.13 (130)	1462 (24)	858 (390)	Most service versions gave 500 hp at 2050 rpm
Benz Bz3 BV	18	8	V, L	200	1800	5.31 (135) × 5.31 (135)	941 (15.4)	683 (310)	
Mercedes type 1468	17(?)	8	I, L	287	1750	5.5 (140) × 6.3 (160)	1201 (19.7)	900 (409)	
Mercedes type 1468	17	6	I, L	180	1500	5.5 (140) × 6.3 (160)	901 (14.8)	700 (318)	Displacement calculated from bore and stroke; has been published as 894
Benz Bz 3a	15	6	I, L	180	1400	5.5 (140) × 7.5 (190)	1070 (17.5)	605 (275)	Bz 3 was 160 hp, 875 cu in.
Benz Bz 4	17	6	I, L	230	1400	5.7 (145) × 7.5 (190)	1148 (18.8)	848 (385)	Called "200-hp Benz"
Maybach	16	6	I, L	200	1400	5.9 (150) × 7.5 (190)	1228 (20)	NA	Zeppelin engine
Austro-Daimler	18	6	I, L	200	1400	5.31 (135) × 6.9 (175)	917 (15)	729 (331)	
Hiero	15	6	I, L	200	1400	5.3 (135) × 7.1 (180)	942 (15.4)	695 (316)	Designer was Hoeronimus, maker Warchalowski, Eissler & Co.
Basse und Selve	17	6	I, L	269–302	1600	6.1 (155) × 7.84 (200)	1381 (22.6)	885 (402)	
Mercedes D-IVa	17	6	I, L	260	1400	6.3 (160) × 7.1 (180)	1326 (21.7)	936 (425)	"Overcompressed" for high altitude

(cont.)

Data for Engines (continued)

Maker's name	Date	Cylinders	Configuration	Horsepower	Revolutions per minute	Bore and stroke, in. (mm)	Displacement, cu in. (L)	Weight, lb (kg)	Remarks
BMW III A	18	6	I, L	180	1410	5.9 (150) × 7.1 (180)	1163 (19)	644 (293)	Alternative to "180-hp Mercedes
BMW IV	21	6	I, L	250	1460	6.3 (160) × 7.5 (190)	1938 (31.8)	NA	
Junkers L-5	23	6	I, L	280–310	1450	6.3 (160) × 7.5 (190)	1399 (23)	711 (323)	
Junkers L-8	NA	6	I, L	440	2000	6.3 (160) × 7.5 (190)	1399 (23)	902 (410)	Complete redesign of L-5 with same bore and stroke
Junkers L-55	28	12	V, L	500–600	1398	6.3 (160) × 7.5 (190)	2798 (46)	1254 (570)	
Junkers L-88	29	12	V, L	700–850	1850	6.3 (160) × 7.5 (190)	2798 (46)	NA	
Renault RE 8 Gd	16	8	V, L	220	1700	4.9 (125) × 5.9 (150)	899 (14.7)	474 (215)	Weight probably wrong; seems too light
Renault RE 8 Gd	17	12	V, L	300	1550	4.9 (125) × 5.9 (150)	1347 (22)	794 (361)	
Lorraine 12Ed	22	12	W, L	450	1900	4.7 (120) × 7.1 (180)	1490 (24.4)	935 (425)	
Lorraine 18Ka	NA	18	W, L	650	1850	4.7 (120) × 7.1 (180)	2235 (36.6)	1232 (560)	
Renault 12MC	26	12	V, L	800	2000	6.3 (160) × 7.1 (180)	2652 (43.5)	1342 (610)	
Mercedes F-2	28	12	V, L	1000	1700	6.5 (165) × 8.3 (210)	3294 (54)	1654 (752)	Used only in "speedboats"
Argus A.S. VI	28	12	V, L	700	1300	6.3 (160) × 7.1 (180)	2652 (43.5)	1168 (531)	Available geared and/or inverted
Fiat A-10	10	4	I, L	100	1400	NA	NA	NA	
Fiat A-12	16	6	I, L	268	1700	6.3 (160) × 7.1 (180)	1326 (21.7)	920 (418)	A-12bis gave 300 hp
Fiat A-14	17	12	V, L	725	1730	6.7 (170) × 8.3 (210)	3488 (57.2)	1740 (791)	
Fiat A-15	19?	12	V, L	430	2415	4.7 (120) × 5.9 (150)	1240 (20.3)	800 (364)	Largely an experimental engine
Fiat A-20	25	12	V, L	430	2200	4.5 (115) × 5.9 (150)	1143 (18.7)	733 (333)	
Fiat A-22	25	12	V, L	750	2100	5.3 (135) × 6.3 (160)	1476 (24.2)	1023 (465)	
Fiat A-25	25	12	V, L	1000	1900	6.7 (170) × 7.9 (200)	3322 (54.4)	1874 (852)	

						Bore × Stroke			
Fiat A.S. 5	29	12	V, L	1050	3300	5.4 (138) × 5.5 (140)	1589 (26)	783 (355)	Schneider Cup engine
Fiat A.S. 6	32	24	V, L	2800	3200	5.4 (138) × 5.5 (140)	3178 (52)	2050 (932)	Doubled A.S. 5 with counterrotating drive shafts
Isotta-Fraschini V-1	11	4	I, L	90	1200	5.1 (130) × 7.1 (180)	583 (9.6)	682 (310)	Dirigible engine
Isotta-Fraschini V-6	17	6	I, L	275	1800	5.5 (140) × 7.1 (180)	1014 (16.6)	620 (282)	Cylinders made in pairs
Isotta-Fraschini V-8	18	6	I, L	300	1800	5.9 (150) × 6.7 (170)	1101 (18)	595 (270)	
Isotta-Fraschini V-10	23	6	I, L	350	1850	6 (153) × 7.1 (180)	1202 (19.7)	726 (330)	
Isotta-Fraschini Asso 200	22	6	I, L	250	1850	5.5 (140) × 6.3 (160)	901 (14.8)	616 (280)	
Isotta-Fraschini Asso 500	25	12	V, L	500	1800	5.5 (140) × 5.9 (150)	1691 (27.7)	957 (435)	950 hp in final version (Asso 750 RC)
Isotta-Fraschini Asso 750	29	18	W, L	820	1700	5.5 (140) × 6.7 (170)	2871 (47.0)	1460 (634)	
Isotta-Fraschini Asso 1000	NA	18	W, L	1100	1750	5.9 (150) × 7.1 (180)	3501 (57.4)	1782 (810)	
Isotta-Fraschini Asso Xl.R.	36	12	V, L	700	2400	5.8 (146) × 6.3 (160)	1963 (32.2)	1135 (516)	Late models gave 835 hp for takeoff
Curtiss OX-5	10	8	V, L	90	1400	4 (102) × 5 (127)	503 (8.24)	320 (145)	Usual installed weight about 400 lb (181 kg)
Curtiss OXX-6	12	8	V, L	100	1400	4.3 (108) × 5 (127)	567 (9.3)	401 (182)	
Hall-Scott A-7	16	4	I, L	100	1400	5 (127) × 7 (178)	550 (9)	410 (186)	Liberty cylinder derived from this engine
Packard 905	17	12	V, L	250	2100	4 (102) × 6 (152)	905 (14.8)	800 (364)	
Packard A-2500	29	12	V, L	750	2000	6.4 (162) × 6.5 (165)	2500 (41)	1210 (550)	Other large Packards as early as 1923
Liberty L-4	18	4	I, L	102	1400	5 (127) × 7 (178)	412 (6.8)	398 (181)	Only two built
Liberty L-6	18	6	I, L	215	NA	5 (127) × 7 (178)	825 (13.8)	550 (250)	Postwar Hall-Scott A-4 and A-5 used Liberty cylinders
Liberty L-8	18	8	V, L	270	1850	5 (127) × 7 (178)	1099 (18)	575 (261)	Experimentally developed 330 hp at 1950 rpm
Liberty L-12	18	12	V, L	400	1800	5 (127) × 7 (128)	1649 (27)	786 (357)	Installed weight at least 900 lb (408 kg)

I—inline; V—V-type; L—liquid-cooled; W—W-type; NA—not available. Open figures—U.S. Customary; figures in parentheses—metric. Weights are dry unless otherwise specified.

FOUR

The engine that was built backward: the rotary.

Precursors

The normal and reasonable way to build a piston engine is to have the engine stationary while the pistons turn the crankshaft. If the engine is an inline or a V-type, there is no other way to design it. If, however, the engine is a radial, it is possible to have a stationary crankshaft and let the engine revolve. Odd as this approach may seem, it has been used with considerable success. An engine built to rotate around a stationary crank is called a rotary, and such powerplants were a major type between 1909 and 1918. During that period, rotaries were unmatched in the medium horsepower class for lightness in relation to power.

The first effective gasoline rotaries[1] were built before the turn of the century by a Hungarian-born New York machine-shop proprietor named Stephen Baltzer. His design objectives were good cooling and the elimination of the flywheel. Some of Baltzer's early engines worked well despite a total absence of cooling fins. Early motors generated relatively little heat and power per explosion, and, running slowly as they did, there was enough time between explosions to make heat dissipation relatively easy. The breeze created by the spinning of the cylinders was all that was needed to do the job.

Slow running also increases the importance of the flywheel. The rotational inertia of a flywheel is needed not only to damp the intervals between powerpulses but also to smooth out the pulses themselves, as the torque created by the explosion's push against the piston varies considerably during the 180° of the power stroke. At high rpm this is not noticeable, but a slow-running engine will vibrate noticeably

---

[1]There had been rotary steam engines earlier.

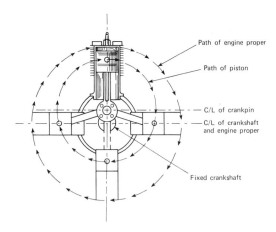

FIG. 4-1   Orbiting path of rotary engine piston.

from this cause alone unless it has a heavy flywheel to smooth it out. A rotary engine solves this problem without the added weight of a flywheel since the whole engine, by spinning, acts as its own flywheel.

In addition, vibration is reduced by the fact, not at first appreciated, that the pistons of a rotary engine do not actually reciprocate. Instead of coming to a stop at the beginning and end of each stroke, they orbit around the stationary crankpin (Fig. 4-1).

Baltzer was successful enough so that he was probably building the lightest gasoline engines for their power in the world by 1899, but this technical success led indirectly to his financial undoing. His work attracted the attention of Charles M. Manly, and he got deeply involved in trying to provide an engine for Professor Langley's human-carrying aircraft. Langley's very successful large model was powered by a rotary built jointly by Baltzer and Manly, but the larger engine for the full-size "Aerodrome" defeated Baltzer. Like many a later designer, he found that a successful design can't always be scaled up. He made promises he couldn't keep, but being a proud and conscientious man, he kept trying until his funds were gone.

Baltzer's engines had originally been built for automobiles, the first of which survives at the Smithsonian. The next rotary of note also had an automotive background. This was the Adams-Farwell, built in Dubuque, Iowa, between 1902 and 1912. Most Adams-Farwells were 5-cylinder en-

gines with longitudinal fins to exploit and accelerate the outward flow of air that accompanies rotating components. Toward the end of the production life of the Adams-Farwell car, the firm built some engines specifically aimed at the aircraft market. At least one of these flew, in a Curtiss pusher, but interest was too slight to sustain production or to justify further development (Fig. 4-2).

The Gnome

The real father of the rotary aircraft engine was a French engineer named Laurent Seguin. His engine was called the Gnome, and it was a fantastic piece of work. The first Gnome had 5 cylinders, in 1908, but the production model was a 50-hp seven. It was soon followed by a nine of 80 hp. The nine then went to 110 hp and, as the powerplant of the early V-strut Nieuports, became a classic fighter engine of the middle war years. A double-row[2] 14-cylinder 160-hp Gnome held the world speed record before the war, and the late 160-hp nine powered the effective but underrated Nieuport 28 in 1918 (Fig. 4-3).

The Gnome employed a peculiar but effec-

[2]Both static and rotary radials have been made in two-row form. A two-row radial is essentially two engines, one behind the other.

FIG. 4-2   An Adams-Farwell rotary aircraft engine, probably of 1911. Adams-Farwell's automobile engines were generally air-cooled. Why did the firm go to water cooling for its aircraft engines? (*Courtesy of the author, from NASM-Smithsonian Hall of Early Flight.*)

FIG. 4-3   110-hp Gnome 9, 1916. A classic fighter engine of the early war years, used in Nieuports, Moranes, and Sopwiths. *(Courtesy of the author, taken at Glenn Curtiss Museum, Hammondsport, New York.)*

tive method of getting fuel into its cylinders. Fuel and air flowed through a hollow crankshaft and thence through automatic intake valves in the piston heads to the combustion chambers. Centrifugal force helped the outward flow of the mixture and, by acting on small counterweights on the valve stems, closed the valves. Thus, unlike the automatic intake valves on stationary engines, those on the Gnome needed no springs to close them (Fig. 4-4).

Since the mixture of gasoline and air filled the crankcase, normal lubrication methods would have failed because the oil would have been washed away by the fuel. The Gnome used the simpler method of introducing a steady stream of oil along with the gas and the air. This meant that oil reached the combustion chambers, where it was partially burned. What didn't burn passed out with the exhaust, and since the only oil that would stand up to the heat and pressure inside a rotary was castor oil, the exhaust was both aromatic and, in its own way, therapeutic.

The exhaust was disposed of as casually as the fuel was ingested. There was no exhaust manifold; the gases were simply allowed to shoot out from the rapidly revolving cylinders. The resulting pinwheel fireworks would have been deleterious to

the structure of a highly flammable airplane, and so a cowling was provided to confine the conflagration. On pusher airplanes these cowlings were usually not used since the fuselage was, so to speak, out of the line of fire. The very considerable improvement in streamlining that resulted was apparently not fully appreciated, though most airframe designers tried to make the fuselage lines fair with the cowling. The uncowled engines used on pushers had a very bad effect on their already dismal performance as a result of the very turbulent airflow reaching the propeller.

The Gnome was also made in "Monosoupape" (single-valve) form. If you are bothered by all those little counterweights bobbing up and down on the regular Gnome, consider the Monosoupape. Surrounding each cylinder were intake ports that were exposed by the piston at the bottom of its stroke. The mixture entered the cylinder through these ports. In operation, the exhaust valve opened before the end of the power stroke, just before the ports were exposed by the descending piston. If all went as M. Seguin intended, the pressure in the cylinder dropped to just equal that in the crankcase by the time the ports were uncovered. The hot exhaust did not enter the ports to ignite the fuel and explode the crankcase. The piston then rose on the exhaust stroke to push out the exhaust gas; the valve closed, and the piston descended on the intake stroke. No fuel was sucked in, because there was no opening for it, until the dropping piston again exposed the intake ports, whereupon fuel entered the cylinder in a rush and was compressed in normal fashion by the compression stroke. It will come as no surprise to learn that the Monosoupape was considered to be a little undependable.

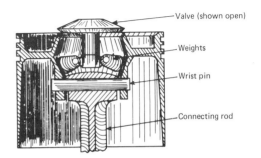

FIG. 4-4   Cross section of Gnome piston, showing intake valve with counterweights.

The Gnome, indeed, every rotary, was beautifully built. The entire engine is made of polished steel, the cylinders have numerous thin cooling fins, there are many small, neat nuts and bolts, and the general design is clean and aesthetically pleasing. The Le Rhone's appearance is enhanced by polished copper intake pipes, and the Bentley has aluminum cylinders. Unfortunately, this excellence of execution did not entirely compensate for the deficiencies of the basic concept. In order to keep the engine light, everything was kept to the minimum possible dimensions. The result was a very limited life. TBO was 50 hr or less. Only one rebore was allowable, and reliability was none too good. There were frequent removals from service for repairs due to wear and breakage before reaching the scheduled overhaul time. Mechanics who had the necessary rapport with these demanding engines were worth their weight in gold to pilots who flew behind them.

Fuel consumption was high to the point where no aircraft intended for long-distance missions was ever powered by rotaries,[3] since the weight penalty of water-cooled engines was more than made up on long flights by their smaller fuel consumption. Performance fell off rapidly as altitude increased, especially for the Gnomes with automatic intake valves. Merely to operate a rotary in flight and in the approach called for special techniques.

In exchange for these headaches the rotary provided performance and maneuverability. Not until the coming of the Fokker D-VII could any fixed-engine machine be expected to dogfight successfully with a rotary-powered pursuit of contemporary design. Its rapid warm-up allowed quick response to an *alerte,* and the absence of a radiator reduced its vulnerability to hostile fire. As long as one-on-one dogfighting was the favored fighting tactic, the rotary reigned supreme; but when the "get in, hit, and get out" tactic became recognized as the better way, the rotary was finished. After 1917 few rotary-powered pursuits could match the straightaway and diving speeds of the better inline-powered fighters.

[3]Though the first crossing of the Mediterranean, by Garros, used a Gnome.

A special problem surfaced as soon as the synchronized machine gun led to combat aerobatics. With 300 lb (136 kg) of engine turning at 1200 rpm in the nose of a 1200-lb (545-kg) airplane, the gyroscopic effect in turns was formidable. A rotary-powered pursuit made instant left turns but turned right reluctantly because gyroscopic procession had to be fought with the controls, and these facts had to be allowed for by both the pilots of rotary-powered planes and their adversaries.

The Le Rhone

The Gnome was built by a firm called Gnome-Rhone, which also made the somewhat different Le Rhone rotary. The Le Rhone had an induction system that was considerably more conventional than the Gnome but, perhaps to equalize the complexities, had a remarkable method of connecting the link rods to the master rod. In the Le Rhone, the fuel mixture went first to an annular chamber at the back of the crankcase and thence, by attractive polished copper pipes, to conventional intake ports and cam-operated valves in the cylinder heads (Fig. 4-5). But where the Gnome used a master rod with link rods, exactly like modern radials, the Le Rhone carried the slipper bearing principle to the ultimate *reductio ad absurdum.* (See the discussion of the Manly-Baltzer engine in

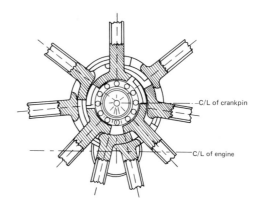

FIG. 4-5   Grooved crankpin and bronze-shoe slipper bearings of 110- to 120-hp Le Rhone rotary engine, 9 cylinders. (Redrawn by author from original catalogue illustration.)

Chap. 2.) Its master rod had three concentric grooves to take slipper bearings from all the other cylinders (it was split to allow assembly). The remaining rods carried bronze shoes, shaped to fit in the grooves, at their inner ends. Counting the master rod as no. 1, the shoes of no's. 2, 5, and 8 rode in the outer groove, those of 3, 6, and 9 in the middle groove, and 4 and 7 in the innermost one. In the picture, no. 1 can be seen in the inner groove, but it did not slide; it was locked in position. While the geometry, as with all slipper bearings, is perfect, mechanically the whole affair is outrageous. Yet thousands of these things were built and they worked well. Even today, one can turn one of these 60-odd-year-old engines over by hand and feel it moving as smooth as silk.

The Le Rhones employed an unusual method of valve actuation. A single rocker arm, pivoted near its center, was made to operate both the exhaust valve and the intake valve. Pulled down, it opened the intake valve; pushed up, it opened the exhaust. To do this, the rocker had to be actuated by a push-pull rod instead of by the usual pushrod. This, in turn, meant that the cam followers had to have a positive action and a system of links and levers accomplished this. This system works well enough—some makers used it up to the late twenties—but its use makes overlap of valve openings impossible. In an engine designed for high power and speed, the intake valve begins to open before the exhaust valve is quite closed, but on the Le Rhone, the rocker arm must clear the exhaust before it can contact the intake. While this puts a limit on power output, it is not necessarily a fault. As it was, most Le Rhone models produced all the power that their structural strength and cooling arrangements could cope with.

Le Rhone 80-hp models were made under license in the United States by a Pennsylvania firm, Union Switch and Signal, and 110-hp models in Sweden by Thulin. Oberursel made the 110-hp model, supposedly without authorization,[4] in Germany. Less ambitious than some, the largest wartime Le Rhone gave only 130 hp. As rotaries went, they were dependable engines.

[4]The Oberursel actually drew on both Le Rhone and Gnome patents but was more Le Rhone than Gnome.

The Clerget and the Bentley

The third French rotary was the Clerget. It had neither the peculiar intake arrangements of the Gnome nor the strange connecting rods of the Le Rhone. Its valves were actuated by conventional rocker arms from two pushrods per cylinder, these constituting a recognition feature. Clergets were made in 110- and 130-hp models, and the 130 may have been a little overdeveloped, since it was subject to overheating. It used a special type of piston ring, called an "obdurator" ring, below the wrist pin to block heat transfer from the combustion area to the lower part of the cylinder. When this ring broke, as it was prone to do, the cylinder turned blue from the heat; a blue color meant that the steel had been heated to 600°F (316°C). Clergets were generally very well engineered except for this problem, those made in England by Gwynne being especially excellent.

This combination of a generally good engine with a heat problem led to the development, from the Clerget, of the ultimate rotary. The British Navy sent Com. Walter Owen Bentley, the same Bentley who built the great sports cars of the twenties, to Flanders to look into the overheating problem on its Clergets. Just before the war, Bentley had improved the performance of the DFP cars he imported by installing some of the first aluminum pistons ever made, and he turned to that metal for the solution to the Clerget's heat problems. Reasoning that the trouble lay in the relatively low heat conductivity of ferrous metal, he designed a new Clerget cylinder comprising a thin steel sleeve in an aluminum shell. This proving successful in Clergets, Bentley next undertook a redesign of that engine. The result was, first, the BR1[5] of 150 hp, then, in 1918, the BR2 of 230 hp (Fig. 4-6).

The smaller Bentleys went into the later Sopwith Camels. A Bentley Camel performed as well as a Clerget Camel but was more dependable. The BR2 went into the Camel's successor, the Snipe. The Snipe was probably the best fighter of the war. An argument could be made for the Nieuport 28, but only the U.S. First Pursuit Group

[5]BR stands for "Bentley Rotary," purely in honor of its designer; there was no Bentley engine factory, all the BR engines being built by contract manufacturers.

FIG. 4-6   Bentley B.R. 1 rotary, the most rational of a somewhat irrational breed. *(Courtesy of the author, taken at NASM-Silver Hill.)*

("the First") was able to tame the 160-hp Gnome Monosoupape that powered this last of the rotary-engined Nieuports.

In France, Nieuports and rotaries went together. The first Nieuport scout, the 11, used the 80-hp Le Rhone, and the last used the big Gnome. All the Nieuports were fine climbers, highly maneuverable, but structurally marginal. By the time America entered the war, the French had largely standardized on Spads and were thus able to spare Nieuports—first old 17s, then new 28s—to equip the American forces. The 28 had been offered to the French, but, with Hisso and Spad production at adequate levels, it was not accepted. Most American units were happy when they got Spads to replace the Nieuports after a few months, but not the First. The First considered the Spad a truck. To them, even a Fokker D-VII was clumsy in comparison with their beloved Nieuports.

German Rotaries

In Germany, the Oberursel rotary kept Anthony Fokker in business. Denied the Mercedes engines he wanted—he cried favoritism, but the fact was

that the better planes built by Albatros and Pfalz had a deserved priority—he bought the Oberursel firm. Rheinhold Platz then designed the Triplane for Fokker, around the 110-hp Oberursel. A Triplane fuselage with cantilever biplane wings made the little-known D-VI which, enlarged to take the Mercedes, became the immortal D-VII. So superior was the D-VII that Fokker's whole factory had to be devoted to its production with the result that the shoe was now on the other foot as far as engines were concerned; he now had all the Mercedes he could use, but there was no market for his Oberursels. This was intolerable to a man as grasping as Fokker, and his response was to let Platz do what he had wanted to do all along: design a cantilever-wing monoplane. The resulting D-VIII, built in only small numbers in late 1918, attained performance on 110 hp that equaled the best that wire-braced biplanes could get from 50 percent more power. This demonstration of the gains achievable through clean and efficient design was ignored by most air forces (the Poles were the exception) for the next 17 years.

The unusual Siemens was the only indigenous German rotary to see combat. To explain what its designers had in mind, the air resistance of the rotary's whirling cylinders must be examined. This resistance absorbed an amount of power that varied as the square of the engine's rpm. As a result, no conventional rotary ever ran at over 1400 rpm, and even the increase from the first rotaries' 1200 to 1400 meant an increase of 36 percent in "windage" drag (as it was called). To get around this problem, the Siemens coupled both the engine and the crankshaft to the engine mount by gears so that the cylinders turned at 800 rpm in one direction and the propeller at 800 in the other. This called for a four-bladed propeller of large diameter to absorb the engine's 160 hp. The Siemens-Schuckert D-IV pursuit, the only plane to use this engine in service, had very long landing gear legs and a steep ground angle as a result of this big propeller. This rare airplane was said to be an even better fighter than was the Fokker D-VII when everything was working well, which, however, was more the exception than the rule. The trouble was that running at 800 rpm didn't generate enough breeze to keep the engine cool. The Siemens seems to have been one of the many engines that passed its official tests but did poorly in service.

Fact and Fiction about the Rotary Engine

Flying behind a rotary would give most modern pilots conniptions. For one thing, rotaries had no carburetors or throttles as such. They didn't need a carburetor to mix the air and the gasoline because the turbulence and heat to which the mixture was subjected as it passed through the engine's interior did the job quite well. All that was necessary was to provide a needle valve for fuel control and a flap for the air. To operate this combination, the engine was first started in a nearly flooded condition. It frequently caught fire from the spilled excess fuel. When this happened, standard procedure was to shut off the gas and let the engine run on the excess while the fire, usually confined to the cowling and the grass under the nose, burned itself out. As the intrepid mechanics dragged the plane backward out of the blaze on the ground, the equally fearless pilot carefully watched for the right instant to re-open the fuel valve. As soon as the engine was warmed up, the air flap was opened wide, the fuel was cut back until the engine started to miss, and the fuel valve was opened until the engine ran smoothly again; that was all there was to it. The engine was left, in general, wide open for takeoff, climb-out, and subsequent flight.[6]

This was all well and good, but there remained the problem of reducing power for landing. The astounding solution, if it really was a solution, called for intermittently cutting the ignition. There were two ways of doing this. The simplest involved what was called a "blip switch," a button on the control stick which, when pushed, grounded the magneto. As the engine's whole weight was revolving, it would continue to turn for some time after the ignition was cut and would pick up again when the blip switch was released. I once saw a Sopwith Pup replica demonstrate how very well this primitive method could work in the hands of an expert. Landing into a 15-mph (24-km/hr) wind, the Pup was held at heights of between 1 and 5 ft (0.3 and

1.5 m) off the grass for about 300 ft (92 m) and was finally allowed to touch down at exactly the spot where the ground handlers stood ready to grab its wing tips.

The "blipping" method should not be considered as the forerunner of the deplorable modern practice of making low and slow approaches and depending on occasional bursts of power to get the aircraft to the airport. Throughout the rotary-engine era and, in fact, right up to the beginning of the present "anyone can fly" period, pilots concerned about longevity made short, steep approaches. This practice was based on a well-grounded suspicion that the engine just might not pick up after cooling off on the approach and that it was better to have too much height on the approach than too little; one could always slip away the excess.

The Clerget and Bentley rotaries had a different way of reducing power for landing. They were equipped with a selector switch by means of which the engine could be made to run on 7, 5, or 3 cylinders instead of the full 9. Whether the engine had a blip switch or a selector switch, the flow of fuel continued undiminished when the engine coasted or slowed down. The unburned fuel came out the exhaust ports and could be counted on to catch fire when the blip switch was released and the engine was started up again (in Gnomes and Le Rhones) or from the hot exhaust of the still-operating cylinders (in Clergets and Bentleys). The pyrotechnic effect when the evening patrol came home must have been worth seeing.

There were legends about the rotaries. One came from the presence of partly burned castor oil in the exhaust. The cowling did not completely prevent the air breathed by the pilot from being polluted with a most powerful laxative. The result was that emergency landings sometimes had to be made for compelling personal needs, despite the bottle of blackberry brandy commonly carried as an antidote. The legend is that some of these landings were perforce in enemy territory and that some pilots became POWs as a result of having to find a bathroom at the wrong time.

Another legend is connected with American mass-production know-how. When we began to build Le Rhones, we decided that the European method of machining cylinders from solid steel bars

---

[6]Some throttling down was possible by adjusting the air and fuel independently.

was ridiculous, and so we undertook to make them from hollow forgings. The tough forgings were harder to machine than were the bars, and so machining time went up, not down. We solved this self-created problem with better steel for the tools and, after only about a year's delay, were able to make just as good cylinders as the Europeans made and do it almost as economically.

Some of the legends are known to be true: for instance, the "Beute" engines. As mentioned earlier, Thulin, in Sweden, was a licensed builder of Le Rhones. The Germans had an engine shortage, their money was good, and the Swedes were willing enough to sell engines. The only problem was that if a German plane with a Swedish engine were shot down in Allied territory, the French might cancel Thulin's license. Taking advantage of the fact that the Germans were known to be using some engines salvaged from downed Allied planes, the Germans worked out a solution. Fake nameplates reading "Beute," meaning "booty" or "loot," were attached to the Thulin engines. It is pleasant to report that the Swedish Le Rhones were of inferior quality and gave considerable trouble.

The legend about the 90-sec alarm response is also well attested to. The U.S. First Pursuit, as mentioned, flew Nieuports powered by big Monosoupape Gnomes. Colonel Hartney's mechanics, many of them ex-race car mechanics from the Midwest, tore down these troublesome engines and drilled holes in the pistons until they looked as though the mice had been at them. This was a common technique of hopping up engines in the days of ferrous metal pistons. It was called "Swiss cheesing" and both lightened the pistons and improved lubrication. Thus equipped, the First was able to forget about warm-up time. It could have a patrol in the air, climbing and in formation, from a cold start, within 90 sec of receiving an *alerte*. Try *that* with a modern powerplant!

By 1918 the end was in sight for the rotary. As long as most engines ran at 1400 rpm, rotaries could have a useful role in combat; but by then there were reliable powerplants running at 2000, and the weight advantage of the rotaries had shrunk to insignificance. In addition, power had risen to levels that no rotary could hope to match. The fine water-cooled Salmson radials were demonstrating that the radial configuration to which the rotary owed its lightness was practical in a stationary engine, and British and Italian work with aluminum air-cooled cylinders had reached the point where it was clear that the whirling of a rotary was no longer necessary for cooling. The British, in fact, thought that they had, in the ABC (All British Corporation) Dragonfly, a stationary radial that was ready for production and service. There were to be no new rotaries after the big Bentley BR2. The future lay in the stationary types.

Rotaries had a short post-war life despite the large number of surplus engines available. For a few years the British and French military used 80-hp Le Rhones in training planes, but their high operating cost made them unsuitable for private flying. In Europe, government subsidies for flying clubs allowed the purchase of newer types with longer life, better fuel economy, and such safety features as dual ignition. In the United States, the ubiquitous OX-5, poor engine that it was, swamped the market at such low prices that the rotaries went begging. Today, of course, a good or rebuildable rotary is worth its weight in dollar bills to enthusiasts who are restoring old planes or building replicas. When one is found, it tends to be in quite good condition, as the castor oil, with which they became coated, oxidized to an effective protective varnish. Cleaned up, they are wonderful to look upon.

Data for Engines

Maker's name	Date	Cylinders	Configuration	Horsepower	Revolutions per minute	Bore and stroke, in. (mm)	Displacement, cu in. (L)	Weight, lb (kg)	Remarks
Adams-Farwell	07	5	Ro, L	NA	NA	NA	NA	NA	
Gnome Lambda	11	7	Ro, A	80	1200	4.9 (124) × 5.5 (140)	773 (12.7)	207 (94)	These Greek-letter names not used in practice
Gnome Delta	14	9	Ro, A	110	1200	4.9 (124) × 5.9 (150)	993 (16.3)	297 (135)	Official rating was 100 hp
Gnome Lambda-Lambda	13	14	Ro, A	160	1200	4.9 (124) × 5.5 (140)	1446 (23.7)	396 (180)	Monosoupape type
Gnome type N	18	9	Ro, A	160	1350	4.5 (115) × 6.7 (170)	970 (15.9)	290 (132)	Monosoupape type; fuel consumption very high
Le Rhone type J	16	9	Ro, A	120	1200	4.4 (112) × 6.7 (170)	920	323	Known as "110-hp Le Rhone"
Clerget type 9	16	9	Ro, A	110	1200	4.7 (120) × 5.9 (170)	931 (15.3)	295 (134)	
Clerget type 9B	17	9	Ro, A	130	1250	4.7 (120) × 6.3 (160)	992 (16.3)	381 (173)	
Bentley BR1	17	9	Ro, A	150	1250	4.7 (120) × 6.7 (170)	1053 (17.3)	405 (184)	Made by several manufacturers
Bentley BR2	18	9	Ro, A	230	1300	5.5 (140) × 7.1 (190)	1522 (24.9)	500 (227)	
Oberursel	16	9	Ro, A	110	NA	4.9 (124) × 5.9 (150)	995 (16.3)	NA	
Siemens-Halske	17	11	Ro, A	160	900	4.9 (124) × 5.5 (140)	1431 (23.5)	428 (195)	Effective speed was 1800 rpm

L—liquid-cooled; A—air-cooled; Ro—rotary; NA—not available. Open figures—U.S. Customary; figures in parentheses—metric. Weights are dry.

65

Cast-block engines.

From the Hispano-Suiza to the Merlin, 1915-1946

Marc Birgikt was a remarkable man. He invented the sports car, built 6-cylinder cars that could go 100 mph (120 km/hr) yet which idled so smoothly that a dime could be balanced on the radiator, made 12-cylinder cars that were better than any Rolls-Royce, and designed the first good aircraft cannon. He also turned out, in 1915, a 150-hp V-8 aircraft engine that showed the world how a water-cooled engine ought to be built.

Birgikt's firm was called Hispano-Suiza—meaning Spanish-Swiss—in recognition of the nationalities of its principal backers and of Birgikt himself. The original factory was in Barcelona, but a Paris branch soon outgrew the Spanish plant. Its name was commonly shortened to Hisso.

The Hisso differed in two ways from all existing aero engines. Instead of having individual cylinders attached to the crankcase, it used aluminum cylinder blocks with screwed-in steel sleeves. This construction saved weight, stiffened the crankcase, and probably led to easier and faster manufacture, once the necessary technique had been mastered. In addition, all its moving parts were enclosed in oil- and dust-tight covers and lubricated from the engine's central pressure-oiling system. It was rigid, light, durable, and dependable. The original Hisso was eventually developed to give 220 hp; a later, enlarged version gave 300. It powered the Spad, the S.E. 5, and a few lesser aircraft, and it was eventually built, not without difficulties, in England and America as well as in France.

Licensed builders found it a difficult engine to manufacture. The aluminum castings made severe demands on the skills of the foundries of the time, and the long fine-threaded sleeves called for precision beyond the capabilities of most machinists. Some makers, such as Brasier in France, who eventually mastered the job of building the basic engine, were unable to cope with the reduction gear drive that Birgikt shortly added to the front end. This gearing was as demanding of nearly perfect metallurgy and heat treating as the rest of the engine was of other techniques. Brasier gears tended to fail after just a few hours' running. Other licensed builders had different troubles. Wolseley and Sunbeam Hissos leaked and broke crankshafts, at least until Wolseley did some redesigning to adapt the engine to its own capabilities. French Mayen-built copies had to be overhauled as soon as delivered before they could be used. A Hispano-built Hisso was a very fine engine, but a licensed builder's Hisso was very apt to be a bad one.

The American licensee was Wright-Martin, a firm formed by the merger of two of the pioneer U.S. aviation companies. Wright-Martin didn't make any bad engines, but that was partly because it didn't make any engines at all during the first year of its license. It had taken up the American rights as early as 1916, even before real production got started in France, but it had a terrible time getting started. It partly solved its problems by derating the Hisso back to its original 150 hp (the production models in France developed 180 hp almost from the start). By late 1917 Wright-Martin was producing good engines of too little power for combat use (some went into JN4-Ds) while European makers were building 180- to 200-hp engines of too poor quality for that same wartime use (Fig. 5-1).[1]

In the course of turning Birgikt's design into

a somewhat lesser engine that it could build successfully, Wright-Martin made one modification that held considerable significance for the future. The original design employed closed-end cylinder sleeves (Fig. 5-2). This meant that the valves could seat directly in openings in the top of the sleeve, but it also meant that the heat of combustion had to travel through the steel, pass the joint, and travel through the aluminum of the head before it was carried away by the cooling water. Wright went to open-ended sleeves and seated the valves against inserts in the head, a change that improved both cooling and ease of manufacture. All Hisso V-8s had rather thin exhaust valves that were somewhat prone to warpage, especially in installations where a short exhaust pipe allowed cold air to be drawn in when the engine was throttled down, but the Wright-Hisso was somewhat better in this respect than the original type.

While Wright was struggling with the prob-

[1]Although, for a while, the British authorities were sending S.E. 5s to the front with very poor Mayen engines, reasoning brilliantly that poor engines were better than no engines at all.

FIG. 5-1   Wright-Hisso model E, 180 hp, 1700 rpm, 470 lb (214 kg), 1920. Direct drive. Compare the clean, neat lines of this engine with the clutter of almost any of the separate-cylinder engines in Chap. 3. *(Courtesy of NASM-Smithsonian.)*

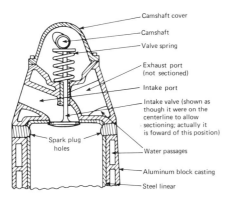

Camshaft cover
Camshaft
Valve spring
Exhaust port (not sectioned)
Intake port
Intake valve (shown as though it were on the centerline to allow sectioning; actually it is foward of this position)
Spark plug holes
Water passages
Aluminum block casting
Steel linear

FIG. 5-2 Cross section of typical Hispano-Suiza cylinder.

lems of producing the Hispano, Curtiss, its great rival, was having troubles of its own. Charles B. Kirkham had been working with Curtiss on an engine that was even more advanced than the Hisso, but Kirkham's engine, unlike Birgikt's, required several years' work before it was ready for service.

The Hispano was a short, bulky engine. This accorded with the then-current idea that a fighter should have its weights concentrated in the interest of quick maneuverability in dogfights. While this bulk did not trouble most designers, Grover Loening and a few others were beginning to stress minimal frontal area as a factor in attaining speed. Kirkham's K-12 design for Curtiss reflected this thinking. His objective, a width no greater than that across a pilot's shoulders, called for a twelve rather than a V-8. Length would have to increase, but cross-sectional area could be reduced. To further reduce bulk, Kirkham abandoned the wet sump that made for a deep oil pan on the Hisso and went to a dry-sump system, as on the German sixes.

Internally, Kirkham introduced two other changes from the Hispano concept, while retaining the essential features of an aluminum block and an enclosed valve gear. His most important innovation was the use of wet, instead of dry, sleeves. By letting the sleeves come into contact with the cooling water, cooling was improved, and manufacturing problems were probably reduced. His other change was somewhat more questionable: he used two camshafts for each block.

To explain this decision, a word about combustion-chamber design is called for. It will be remembered that good breathing is essential for high output, and this requirement puts a premium on the use of as much valve area as possible. If an engine has a flattop combustion chamber and two valves, the size of the valves is automatically limited. The simplest solution is to enlarge the bore to accommodate bigger valves while simultaneously reducing the stroke to maintain the desired piston displacement, but this solution, while elegant, is not conducive to good bearing life in an inline engine that operates under load at a relatively slow speed. The alternative, if the use of two valves is insisted on, is to slope the valves and use some form of hemispherical head. This gives an efficient shape to the combustion chamber but does tend toward more frontal area. Kirkham, and most designers of V-12s since his time, used four valves per cylinder instead. This gives the needed valve area and has the further advantage that the exhaust valves are smaller and easier to keep cool. Unfortunately, simplicity is lost (Fig. 5-3). When two valves per cylinder are used, they line up in a row and can be operated by a single overhead camshaft, as was done on the Hisso. Four valves per cylinder make two rows, and a single camshaft can't do the job unless rocker arms are used. Kirkham's decision was to use two camshafts. Although most later V-12s used rocker arms, the two pairs of camshafts on the K-12 were never a source of trouble. Plenty of other things were, though.

Just how advanced this engine was can be seen from a comparison with the Liberty, designed over a year later. Both were 400-hp engines, but

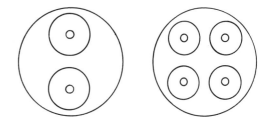

FIG. 5-3 How four valves per cylinder provide more valve area and better breathing in a flat-topped cylinder.

the Liberty got its power from 1650 cu in. (27 L), whereas the K-12 needed only 1145 cu in. (18.76 L). The K-12 operated at as much as 2500 rpm, compared with the Liberty's 1800, and was able—more or less, as it turned out—to take this high speed because its *en bloc* construction made it far more rigid than the relatively limber Liberty. In a Curtiss Triplane, also designed by Kirkham, the K-12 set world speed and altitude records; yet it was a decidedly faulty engine.

It had three main problems. One came from the use of a one-piece block. Had this worked, a fine combination of strength, lightness, and freedom from gasket problems would have resulted, but such a casting was beyond the capability of any foundry. It was not until 1935, with the Lincoln-Zephyr car, that a 12-cylinder block was successfully cast in one piece, and the Zephyr block was in iron. As far as I know, nobody ever cast a large 12-cylinder block in aluminum. As a result, most blocks reaching Curtiss had to be rejected.

A second problem was that the use of a four-bearing crankshaft led to poor bearing life. This was a design gamble that, had it worked, would have allowed the engine to be shorter than it would have been had the more conservative seven bearings been used.

The third problem lay with the reduction gearing. This was a necessity for a 2500-rpm engine. Even setting aside the low efficiency, because of supersonic tip speeds, of a propeller large enough for 400 hp at 2500 rpm, no wooden propeller of the time could have stood up to such a speed. Unfortunately, the design and construction of a suitable gearbox defeated Kirkham and the Curtiss organization.

Kirkham left Curtiss amid disagreement over the question of how much redesign the K-12 needed, but not before having built the K-6 and the C-6, each being in effect half a K-12. These were nice little direct-drive engines delivering 150 hp but were not needed at that stage of the rearmament program (Fig. 5-4). The design then went through a series of changes, described in Hugo Byttebier's fascinating Smithsonian monograph on the engine, and emerged, completely

FIG. 5-4 Curtiss K-6 engine, 150 hp. Note the tiny electric starter with double reduction gearing to enable it to turn over 573 cu in. (9.39 L). The large intake manifold is a sort of plenum, with relatively low velocity and relatively high pressure for the fuel mixture, to promote even fuel distribution to the cylinders. *(Courtesy of the author, taken at Glenn Curtiss Museum, Hammondsport, New York.)*

redesigned by Arthur Nutt, as the direct-drive D-12 of 350 hp. Just as with the Baltzer-Manly, the Clerget, the Bentley, and the Whirlwind (see Chap. 6), the first begetter of the engine needed to have someone else bring it to its full development.

Soon the Reed, the first successful metal propeller (which could use thinner sections than a wooden one), allowed the rpm to be raised back almost to the original level without loss of efficiency. The seven-main-bearing D-12, now with separate blocks and crankcase, gave 425 hp at 2250 rpm in racing versions by 1922. In a Curtiss racer, the D-12 became the first engine ever to power a plane to 200 mph (322 km/hr), and during the first half of the twenties, D-12-powered Curtisses swapped the unlimited record back and forth with Hisso-powered Spads and Nieuports in France. It won two Schneider Cups and was a threat at the National Air Races as late as the mid-thirties. It consistently beat planes powered by Packards that had up to twice the D-12's displacement.

The D-12 was as effective as a workhorse as it was as a racer. Its durability and reliability enabled Curtiss to sell it to the Army on the legitimate grounds that a newly purchased D-12 was cheaper in the long run than a Liberty obtained free from surplus stock. It powered all the Army's fighters until the coming of the Pratt and Whitney Wasp. The Curtiss Falcon observation plane, which replaced the D.H. 4 and served well into the thirties, used it in numbers. Beyond doubt, and with all respect to the Napier Lion, it was the world's outstanding engine in its class until at least 1927.

Up to the mid-twenties the only British cast-block aero engine to be produced in quantity had been the BHP,[2] an enlarged and redesigned 160-hp Beardmore. Its design and manufacturing history is complicated. All sorts of programs were organized around it in the belief, formulated before it was made to prove itself, that it was going to be a wonder engine. The good D.H. 4 was redesigned as the poor D.H. 9 to use it, and two firms were to produce it—Galloway Engineering, a Beardmore offshoot, and Siddeley-Deasy Motors.

At first, Siddeley was to work on the as-

designed aluminum-block version while Galloway was to start on an iron-block modification as insurance. The Puma, as the Siddeley product was called, reached production before it became necessary to fall back on the iron-block version. In a way, this was too bad, because the iron-block model would have made a good anchor, giving at least *some* positive results from the program. In general—although the Puma wasn't as bad as the Adriatic, as the Galloway version was called—the BHP was pretty much a dog.

Its 230-hp output was less than expected, and it was troublesome in service. At 1148-cu in. (18.82-L) displacement, it was an awfully big engine for its weight of 680 lb (308 kg) and may have been structurally fragile. The Germans and the Italians built even bigger sixes than the BHP, but their designs were better, having smaller bores and longer strokes as well as, apparently, more structural strength.

The BHP did not look like most other aluminum-block engines in that the water jackets went only part way down the cylinders, just as they had on its welded-cylinder parent, and in having detachable heads bolted to flanges on the cylinder block.

It stayed in service for a few years after the war, but apparently the RAF (Royal Air Force) was glad to see the last of it. Experience with the BHP did not make the RAF a cast-block enthusiast, and so much credit is due it for having the percipience to see that something along D-12 lines was what it needed (Fig. 5-5).

In 1924, Richard Fairey, head of one of the largest British aircraft firms, took out a license to build the D-12 in Britain. At the time the two British engines in the 400-hp class were the Rolls-Royce Eagle and the Napier Lion. Both were fine engines but were heavier and bulkier than the D-12. When used in "general purpose" two-seaters loaded down with all the goodies thought necessary by a peacetime military establishment, the resulting performance was below that of the better planes of 1917. Fairey, disgusted with building this sort of plane, designed around the D-12 a light bomber, called the Fairey Fox, that could run away from the fighters of the time. The RAF bought a few Foxes—it was so advanced an airplane that the RAF had little choice but to do so—but the main

[2]BHP stood for Beardmore (the prime contractor), Halford (the designer), and Pullinger (added to make the initials the same as for brake horse power).

FIG. 5-5   230-hp BHP engine. The first large-production British cast-block aero engine. The somewhat un-aeronautical appearance, especially the flange-mount-ed cylinder heads, may reflect the fact that Beardmore had been a marine engine builder. The BHP, had it been a bit larger, would have looked more at home in a tugboat's engine room than in an airplane. *(Courtesy of NASM-Smithsonian.)*

repercussions of Fairey's enterprise were felt in the engine industry.

Because civil aviation in Britain and America was becoming heavily oriented toward air-cooled radials, the specialists in liquid cooling were in a position of having to jump when the military said "Frog!" The British military told Rolls and Napier to get cracking on a domestic equivalent of the D-12. The RAF was not about to support with orders three builders of the same class of engines. Eliminating the D-12, which was to have been called the Fairey Felix, was easy. All that had to be done was to subject it to the official RAF acceptance test. It will be no surprise to those acquainted with government testing that this engine, the most reliable of its era, failed the test.[3]

Government testing of aircraft and engines is everywhere a peculiar business. It is as nearly irrelevant to reality as anything could possibly be.

[3]Later, when the Kestrel project had gotten under way, the D-12 passed the official test.

Not only do good engines often fail, but consider the partial list of bad engines that have passed. In the United States we had the Packards, gas and Diesel; the Poyer; the high-temperature Conqueror; the Caminez; and the Aircat. The British passed the BHP, the ABCs, the big Beardmore six, and various barrel engines. Germany accepted the big Siemens. The Japanese approved the Nakajima Homare and the license-built DB-603, and the French accepted the Hisso radials. This applies to airframes as well. At the time this was written, the U.S. government had just had to admit that its testing of the DC-10 was probably inadequate. The Martin 2-0-2 satisfied the government until its wings began to fall off from metal fatigue before 2 years' service had been completed. So did the DC-6's fire extinguishing system that later asphyxiated pilots in flight. Yet the perdurable DC-3 flies today on waivers because it doesn't meet present-day standards for—of all things—spin recovery when loaded to maximum aft center of gravity limit! The classic case may be the British approval of the Comet I, but there are many others. The trouble, at least on the surface, seems to be that any government department would rather spend a dollar on simulation than a dime on in-service testing, and the simulation frequently misses vital points while stressing irrelevancies.

Napier bowed out of trying to make an engine similar to the D-12. An old firm, it seems to have been suffering from a near-terminal case of nepotism. Rolls took up the challenge and came up with a splendid engine, called at first the F and, later, the Kestrel. Its designer was Arthur Rowledge, one of those rare men who have been able to make themselves into world-class designers despite the deficiencies of British technical education. Rowledge had been the designer of the Napier Lion, but he had got a bellyful of Napier management and had moved to a place where his great gifts were appreciated and utilized.

Rowledge took advantage of all that had been learned since the D-12 had been developed. The Kestrel, though not much larger than the D-12 [its displacement was 1296 cu in. (21.24 L)], was sturdy enough to possess more potential for development than its model had. It came out in 1927 at 450 hp and went to 750, at 3000 rpm, before it was retired; a 1939 revival under the name "Peregrine" gave 885. Like the original K-12, the Kestrel relied on reduction gearing to allow high speed, but Rolls was far more successful than Curtiss was at building durable gearsets. The gear case and, indeed, the whole engine looked as though ruggedness had been put ahead of lightness throughout the design process. The Kestrel weighed about 865 lb (393 kg), compared with the D-12's 680 lb (309 kg). Some of this additional weight went into a supercharger and its drive mechanism, whereas the D-12 went unsupercharged (or what came to be called "naturally aspirated") throughout its production life.

The D-12 possessed one retrograde design element in its use of the old-style closed-end cylinder sleeves. On the Kestrel the valves seated in inserts in the head, which, unlike that on the D-12, was in one piece with the block. There were four valves per cylinder, as in the D-12, but the Kestrel used a single camshaft and rocker arms. The exhaust valves had hollow stems partly filled with sodium, which sloshed back and forth for better

heat dissipation, this being one of the first applications of an improvement that soon became universal (Fig. 5-6).

The best-remembered airplane to use the Kestrel was the Hawker Fury, fastest and most beautiful of all open-cockpit fighters[4] and the first service type to exceed 200 mph (325 km/hr) with full military load (Fig. 5-7). But, like the D-12, the Kestrel was also quite at home in slower planes. The RAF employed it in bombers and patrol planes whose top speeds were from 50 to 75 mph (62 to 93 km/hr) slower than the Fury's, and which required hours of continuous running, with entire satisfaction.

The excellence of the Kestrel set the RAF firmly on the road to its World War II commitment to liquid cooling. The Kestrel was, in addition, a key element in an experiment that had great significance for two nations in that war. Rolls, needing the fastest plane available for experiments with new radiator designs, bought a German He (Heinkel) 70 in 1933. This was a low-wing, retractable-gear mail plane, capable of seating six passengers. It went 234 mph (380 km/hr) with a bulky 600-hp BMW engine and an unbelievable 260 mph (423 km/hr) behind a Kestrel. This was a stunning 35 mph (57 km/hr) faster than even the cleaned-up Super Fury with the same engine.[5] The lesson it taught could hardly be ignored; the biplane fighter was finished. Within a year the Air Ministry had issued its first specification for a retractable-gear monoplane fighter, and the path to the Hurricane and the Spitfire was open.

The same international hybrid had its effects in Germany. Just as the He 70's performance opened British eyes to the obsolescence of the biplane fighter, so did the qualities of the Kestrel demonstrate to the Germans that the welded-cylinder construction of the BMW had reached the end of its useful life. Both the Spitfire and the D-B series of engines owed their existence in large part to the Kestrel installation in the He 70.

[4]Those familiar with Continental aircraft may challenge this statement on behalf of the Czech Avia 534, but the Avia was obsolete almost as soon as it reached service status in 1934, whereas the Fury was very advanced for its date. Besides, most Avias had cockpit canopies.
[5]Some sources say the Heinkel had an early Merlin. Here I am following Heinkel's autobiography.

FIG. 5-6(a) Early, unsupercharged Rolls-Royce Kestrel. Actually, the early, naturally aspirated Kestrels were called simply F. Most production Kestrels were supercharged. *(Courtesy of Rolls-Royce Limited.)*

Meanwhile, back home, the D-12 had another offspring, the Conqueror. Announced in 1928, it incorporated many of the advances that had gone into the Kestrel. With a displacement of 1570 cu in. (25.7 L), it had a reliable gearset, open-ended cylinder sleeves, less weight than the Kestrel, and frontal area hardly greater than that of the D-12. It powered a number of significant aircraft,

notably the striking Curtiss P-6E Hawk, yet it was out of production by 1934 (Fig. 5-8).

Since numerous marks of the somewhat similar Merlin gave 1.05 hp/cu in. (62 hp/L), the Conqueror may have had a potential as high as 1650 hp. As this is more than was developed by any widely used mark of the only liquid-cooled U.S. engine used in World War II, the Allison, it is

FIG. 5-6(b) Rolls-Royce Kestrel. Notice the clutter at the rear. It took more than the use of a cast-block design to overcome Rolls-Royce's distaste for what one British writer called, in connection with an Italian engine, "the Latin passion for superficial neatness." *(Courtesy of Rolls-Royce Limited.)*

FIG. 5-7 Hawker Fury, Kestrel-powered. Note that the radiator is somewhat smaller than those on the somewhat similar Curtiss types. Is this because of the cooler English climate or did Rolls know something about radiator design that Curtiss did not? *(Courtesy of the author, photo of a model.)*

legitimate to wonder whether the United States prematurely abandoned a design that could have made a real contribution in that conflict. If so, the blame must be divided between the Army, the Depression, and Curtiss-Wright. Since water boils at 212°F (100°C), its use as a coolant requires a radiator large enough to bring the circulating coolant down to a temperature well below this figure. Adding ethylene glycol to the coolant raises the boiling point in proportion to the amount added. Since the coolant can be run hotter, a smaller radiator can be used, and the engine can be operated at what is actually a more efficient temperature. Unfortunately, the Army went overboard, demanding a coolant temperature of 300°F (149°C). At this heat the V-1570 (the Conqueror's official designation) turned from a reliable engine into a temperamental one, liable to burn valves at any time without warning.

Possibly something could have been worked out, but Depression economics intervened. The

FIG. 5-8 Supercharged Curtiss V-1570 Conqueror, 650 hp, direct drive, about 1932. Note the way the water pump drive was integrated with that for the supercharger. In all installations, the radiator was below the engine, making a low location for the water pump desirable. (This engine carried serial number 632, which raises an interesting question. If 632 Conquerors had been made by 1932, production must have been at the high rate of 150 to 200 per year and, as the Army Air Corps had less than 300 planes in 1934, they must have been wiping them out as fast as they were built. On the other hand, if this is a continuation of D-12 numbering, 632 seems a rather small total of both types of engines to have been built over 12 years.) *(Courtesy of NASM-Smithsonian.)*

Conqueror was not an inexpensive engine, and the Army had limited funds, which would buy more airplanes if the airplanes were powered by less expensive radials. The Army, therefore, considering the unreliability of the Conqueror at its self-imposed 300°F operating temperature, made a total changeover to radial power (which the Navy, for not quite the same reasons, had already done). The Curtiss A-8 and A-12 Shrikes demonstrated the performance difference when a radial was substituted for a Conqueror in a given airframe. As originally built, the Conqueror-powered A-8 went 183 mph (300 km/hr). With a Wright Cyclone substituted, it became the 177-mph (288-km/hr) A-12. Ceiling dropped by 3000 ft (914 m), despite the fact that the Cyclone was lighter (the Conqueror's supercharger was more effective), and range went up by 85 mi (138 km). This tradeoff was acceptable, and the Air Corps informed Curtiss, in 1934, that it should not count on getting any further military orders for the Conqueror.

If we compare Curtiss-Wright with Rolls-Royce, we find that Curtiss was less technically authoritative and less committed to liquid cooling. Rolls could inform the RAF that the Kestrel and, later, the Merlin were to be run at 260°F (127°C), and the RAF would accept this as binding, but Curtiss's word did not carry anything like the same authority with the technical people at Wright Field. Rolls built only the Kestrel and the Merlin; Curtiss would just as soon sell Cyclones as Conquerors.

As a maker of planes as well as engines, Curtiss-Wright could and did try to develop a civilian market for the Conqueror, but its effort was halfhearted. The firm was in production with a big bomber, the 90-ft (27.4-m) span Condor, to which it fitted an 18-passenger fuselage and which it tried to market as a transport. It looked like something from another age and flew like one, bobbling around in the slightest turbulence on its lightly loaded wings and with its biplane tail twisting and weaving disconcertingly. When the Condor proved a flop in the transport market, Curtiss-Wright got out of the liquid-cooled engine business for good.

The Kestrel itself had two descendants of major importance. One was the big Rolls-Royce R that replaced the bulky Lion in Supermarine's Schneider Cup racers. This was one of the few engines that used supercharging to raise its sea-level horsepower; on special racing fuel it gave an unprecedented 2500 hp. In the Supermarine S-6B it won permanent possession of the Schneider Cup and set the first-ever record of over 400 mph (644 km/hr). The R was not an off-the-drawing-board success. Debugging it put to a severe test the legendary talents of Rolls technicians for turning doubtful prototypes into fine engines—a talent that stood Britain in good stead during the Merlin's long history (Fig. 5-9).

The Merlin was, of course, the Kestrel's other child. Fundamentally it was simply an upsizing of the Kestrel to 1650 cu in. (27 L). It proved to be an engine that seemed to have no limit to what could be done with it. Originally intended to produce 750 hp, it was already giving 950 when the first Hurricane flew in November 1935 and 1050 for the first Spitfire in 1936. Early wartime models were rated at 1175 hp, and by late in the war it was giving a dependable 1660. Some models went to 1750, and there was even a 2000-hp mark. In the United States, Packard built it with detachable heads, a modification soon adopted, to the benefit of the engine, by Rolls itself. The Packard Merlins were the engines that transformed the P-51 from the mediocre fighter that it was (with an Allison up front) into what may have been the best all-around fighter of the war; it was certainly the best long-range one [Fig. 5-10(a) and (b)].

Rolls plugged away at geared supercharger design until it got excellent boost at very high altitudes without absorbing a great deal of power. By the end of the war, there were few organizations that knew more about efficient centrifugal blower design than Rolls; this was knowledge that stood it in good stead when the turbine era began. The only area in which Rolls might be said to have lagged was in fuel injection (it still does in automobiles; the carbureted Rolls had the worst gas mileage of any car tested by the U.S. government in 1979). Even in 1940, Spitfire pilots found that a fuel-injected Me 109 could escape them in a dive because of the hesitation that the carbureted Merlin displayed in a nose-over, yet to the end, all Merlins except one low-production model for the Fleet Air Arm had carburetors (though eventually they were modified to largely eliminate the cutting-out problem).

In addition to its use in fighters, the Merlin

FIG. 5-9   Rolls-Royce R for Supermarine Schneider Cup seaplanes, up to 2500 hp, 1929 to 1931. Note the very large supercharger fed from a carburetor, which in turn had a forward-facing ram air intake. The camshafts are driven from the rear and the camshaft covers are tapered forward to aid in streamlining the nose of the aircraft. Running with pure castor oil in the crankcase, this engine was good for about 1 hour's running at full power, yet 10 years later the Griffon, having the same bore and stroke, was giving up to 2374 hp and normal durability in military planes. *(Courtesy of Rolls-Royce Limited.)*

served Britain well in Lancaster, Wellington, and Halifax bombers, but, in retrospect, it seems that radials, had they been available in sufficient quantity, might have done a better job. Just after the war, Canadair built some DC-4s under license as the North Star, with Merlin engines. While the North Star had a good safety record, the Merlins were not as satisfactory as the Pratt and Whitney R-2000s in the Douglas-built version. Maintenance costs were higher, TBO lower, and in-flight failures more frequent than for the radials. Also, passengers found them noisier, though this has no bearing on their suitability for bombers.

The fact is that a fully developed liquid-cooled engine can be made to develop more power than is, in a sense, good for it. Unless an engine becomes obsolete before it reaches its final development, as happened with the D-12, it tends to go through three stages. In the first, the engine, being a new design, is advanced but unreliable. In the second stage, the bugs have been worked out, power is up, and everybody loves it. In the final stage, it has been developed to give the maximum power of which the basic design is capable, but in the process it has become temperamental. It needs

mechanics and pilots who know it well and can cater to its whims. It seems to be easier to take a liquid-cooled engine to the third stage than an air-cooled one, perhaps because the limitations on heat dissipation are more basic in air-cooled engines. In any case, the last military Merlins, at close to 2000 hp, were decidedly touchy engines. Today the enthusiasts who fly restored World War II fighter planes find the lowly Allison a much more tractable engine than the great Merlin.

France—Hispano-Suiza, Farman, Lorraine, and two *Incroyables*

While these American and British engines were evolving, the French powerplant that had started it all was moving with the times. Hispano-Suiza first scaled up its classic V-8 to 300 hp. The upsizing exacerbated the vibration that had been only a minor fault in the smaller version. Any scaled-up engine is apt to exhibit increased vibration because the relationship between the burning time of the fuel and the length of the piston stroke changes, but part of the Hisso's problem lay with the use of a so-called flat crank. To run smoothly, a V-8

FIG. 5-10(a) Rolls-Royce Merlin III, about 1050 hp. This is an actual Battle of
Britain engine, removed from an aircraft after extensive service.

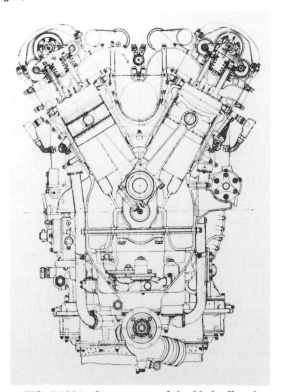

FIG. 5-10(b) Cross section of the Merlin II and
III. Compare with Allison cross section, Fig. 5-18.
(Courtesy of Rolls-Royce Limited.)

should have its crank throws 90° apart, but the Hisso's were at 180°, exactly like those of an inline four. In addition, the big Hisso suffered from an increase in frontal area. It was a reasonably efficient and reliable engine, and the French military used it for several years, but these drawbacks naturally directed attention to the advantages of using 12 cylinders.

The 300-hp Hispano adopted the same advanced form of cylinder construction as the Wright-Hispano, with open-ended cylinder sleeves. The firm thus had two cylinder designs in production, the older being called the "a" cylinder and the newer the "b" type. On these design bases a number of different engines were built. Each cylinder type became available in four kinds of engines: an inline six, a V-8, a V-12, and a W-12. Most were available with a choice of geared or direct drive. There was a wide range of horsepower ratings. The sixes could be had at 150 or 300 hp, the eights at 175 to 300, the V-12s at 390 to 600, and the W-12s at 535 to 635 by 1930, the year when the permutations were all worked out. As all this was in addition to the automobiles described earlier, the firm's technical energy was remarkably at variance with the lassitude that it, in common with so much of the French aeronautical industry, displayed during the late thirties.

In 1931 a line of W-18s was added. This was a natural evolution, since it was largely a matter of using three 6-cylinder blocks in much the same way that three 4-cylinder blocks had been used to make the W-12s. The W-18s differed from the W-12s in that the angle between blocks was 60° on the W-18s instead of 45°. This made for a wide engine—62.36 in. (1584 mm)—but the individual blocks were narrow and could be nicely cowled. Normal output of the W-18s went up to 1125 hp— probably the world's most powerful production engine at the time—and there was a racing version that gave 1680 hp at the remarkably low specific consumption of 0.46 lb (0.209 kg).

With such a wide range of overlapping power ratings available, it is impossible to tell what sort of Hisso powered a given airplane solely by reference to the engine power cited. Hissos never received type names, since both the firm and the government used a systematic but complex coding of

letters and numbers. In all cases, the designation begins with a number representing the number of cylinders and, at least in earlier engines, ends with a lowercase letter for the "a" or "b" type of cylinder, but the rest of the code varied from time to time. For example, by 1930 the subscript letter identified the engine as high or low compression. An 8Fb, for instance, was a large-cylinder V-8 of the twenties, but a 12Ga was a later large-cylinder W-12.

As the thirties moved toward the Second World War, Hispano production and development became concentrated on the 12X and 12Y series, both introduced in 1934. The X line had a displacement of 1636 cu in. (26.8 L), used fork-and-blade connecting rods, and developed just under 700 hp. Weighing 850 lb (386 kg), the X corresponded roughly with the Curtiss Conqueror, though its displacement was slightly larger. The Y, at 2181 cu in. (35.74 L), was a larger engine, used the old-type main-and-link connecting rods, and developed something under 1000 hp, not an impressive figure for a liquid-cooled engine of its size. By this time Hisso engine designations were becoming somewhat arcane. For example, a 12Ydrs had 12 cylinders, clockwise rotation, moderate supercharge, and large displacement (Fig. 5-11).

Few, if any, engines were licensed more widely for foreign production than the late Hispano-Suizas. There were licensees in Great Britain, Spain (Hispano-Suiza's own Spanish branch, soon to disappear in the Falangist disaster), Japan, Switzerland, Soviet Russia, and Czechoslovakia. Soviet Russia did the most with the basic design, ultimately getting 1600 hp from a 12Y derivative. A strange situation that developed in connection with the Czech licensee will serve to illustrate the chaotic production situation in France in 1939. France itself could not produce Hisso engines as fast as it could the aircraft, principally Morane-Saulnier M-S 406 and Dewoitine D-520 fighters, that used these engines. In its extremity, France turned to Skoda, which was able to supply France with 25 engines per month during much of 1939. The Swiss licensee, Saurer, also sold a few engines to the French.

Hispano was testing a 1300-hp version of the 12Y by 1938, but it never reached production until after the war. Not one French fighter ever had

FIG. 5-11 Hispano-Suiza type 12y, 650 hp, about 1936. *(Courtesy of NASM-Smithsonian.)*

a Hisso of over 910 hp, with the result that adequate speed could only be had at the expense of firepower. While the 1200-hp British engines could power a fighter carrying the weight of eight .303 caliber machine guns and the Americans could use six .50s, typical French armament was four .30s or one slow-firing cannon and two .30s. When one reads of the French buying the P-36, with its mediocre performance, from the United States in 1939, any feeling that this represented poor judgment must be modified in the light of the inadequate performance of France's own fighters, an inadequacy resulting from the absence of an engine of sufficient power. The fact is that in 1939 the P-36 was as good a fighter as anything France had.

The French failure to develop the big Hisso to its full potential can be best appreciated in the context of Russian achievements (see p. 79 and 92). Exactly why no French version of this engine reached the necessary power for wartime effectiveness must remain a mystery, but it is hard to believe that purely engineering reasons can explain it.

Politics and ideology led to strange things in prewar France, and it must be in these areas that the explanation is to be found. Eventually, in 1951, a 1500-hp Hisso was made—the 12Z-11Y—to sell as a replacement engine to smaller countries that were still operating World War II planes acquired from the former combatants. It should have been built 12 years earlier.

The Farman firm, well known as an airframe builder, also made aluminum-block engines. It built engines as early as 1916 but did not become prominent until the twenties. No Farman was ever a completely conventional engine. Its first large engine, a 500-hp W-12 called the 12 WE, was unique in that it combined aluminum blocks with pushrod valve actuation. The ends of the rocker arms protruded through holes in the valve covers to engage the exposed pushrods, and so the covers were mere dust shields; the valve gear could not be considered enclosed like that in the Hisso and other engines.

Later in the twenties, Farman built the only

water-cooled inverted V-8 that I know of, its 8 VI, and an inverted 18-cylinder W-type of 730 hp, its 18 W.I. Possibly the firm's urge to be different was satisfied by the inverted configuration, because these engines had conventional overhead (or "underhead") camshafts. These were followed, in 1932, by two inverted twelves, the 12 W.I.—an overhead-cam replacement for the 12 WE—and the smaller 450-hp 12 Brs. The 12 Brs's cylinders were unique, having water cooling for the outer part and air cooling for the inner. This makes a certain amount of sense, since the heat is more intense nearer the head, but I know of no other engine that was made this way. It was an odd size, at 471 cu in. (8 L); few such small twelves were made. This displacement was used because the 12 Brs was intended for the Coupe Deutsch de la Muerthe, Europe's most important air race. Its high specific output is explained by its being a racing engine. Though ingenious, a twelve of this size had no commercial potential.

Possibly the same can be said of the Farman 18T, the only T-type aircraft engine ever built. It had three banks of cylinders set 90° apart from each other (Fig. 5-12). This put two banks in the same position as those of an opposed engine and the other below the crankcase as in an inverted inline. This construction resulted in the excessive width of 54 in. (1371 mm), an obstacle to streamlining unless the intent was to use the engine in a midwing monoplane and fair the outside banks into the wing roots. Perhaps just such an installation was in the designers' minds, since it had been intended for the 1931 Schneider Cup race. It was not ready in time, but, had it been, it would have stood no chance against the Supermarine's 2500-hp Rolls-Royce power. Nevertheless this was an intriguing engine. It was very light for its size, at 1062 lb (483 kg) and 1498 cu in. (24.5 L), and developed approximately 1 hp/cu in. (62 hp/L). Had it been possible to bring it to a point of adequate durability and to get something like the same power on normal gasoline instead of on racing fuel, this might have been a better fighter engine than the big, lazy Hisso.

Farman's liking for inverted engines may have resulted from the type of reduction gears that it used. In a direct-drive engine the advantage of inverted construction comes from the high location of the propeller shaft. In a conventionally geared

FIG. 5-12 Farman 18T, 1200 hp, 1930 to 1931. The only T-type liquid-cooled engine ever built. (*Courtesy of NASM-Smithsonian.*)

V-type, though, the offset resulting from the use of spur gears puts the propeller shaft above the crankshaft if the engine is upright and below it if the engine is inverted. Hence there is little point in putting up with the oil distribution problems of an inverted engine. Farman, however, used epicyclic rather than spur-type reduction gears, and, since this so-called Farman-type gearing results in the propeller shaft being concentric with the crankshaft, some gain did result from the use of the inverted configuration (Fig. 5-13).

Farman engines were apparently used by only two airframe builders, Farman itself and Latècoere. Even Farman airplanes could be had with alternate powerplants if desired. Nevertheless they must have been first-class engines, because they were used in very demanding services. Farman's transport planes flew all over Europe and the Mediterranean, and Latècoeres were used on the South Atlantic and South American routes and for long-range naval patrol work.

Lorraine also built aluminum-block engines. This firm's separate-cylinder engines had competed rather effectively with Hispano and Renault during the twenties, but by the end of the decade the firm must have seen that the end was in sight for the welded cylinder. Its resulting line of cast-block

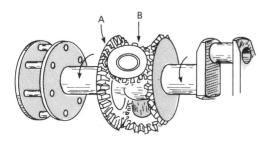

FIG. 5-13 Farman gear set. The stub shaft at right is coupled to the crankshaft and turns the large bevel gear. The spider carrying the four small bevel gears is attached to, and turns, the propeller shaft on the left. "A" is another large bevel gear, identical with the driving gear, which is inside the case and which does not revolve. Attention homebuilders who are looking for reduction gearing for Volkswagen and Corvair engines—the resemblance to an automobile differential is intriguing, but the bearings on the spider gears of an automobile differential are not made for continuous operation under full load!

engines included a 600-hp W-12, the Courlis; a 500-hp V-12, the Petrel; and the big, lazy Eider of 1050 hp from 2743 cu in. (45 L). Neither these nor the firm's concurrent line of medium-size radials were successful enough to keep the company going through the Depression.

Two strange French cast-block engines, both derived from automotive practice, should be mentioned. First in point of time was the Bugatti H-16 of 1918. It consisted of two straight eights, placed side by side on a common crankcase. Both crankshafts were geared to a central propeller shaft that was made hollow to allow a small cannon to be fired through it, a la Hispano. It was built exactly like Bugatti automobile engines. Each half had two cast-iron 4-cylinder blocks with integral heads, and the crankshafts consisted of a four-throw flat crank in front and another, rotated 90° from the first, in the rear. The crankcase, gear case, cam covers, and water jacket enclosing plates were aluminum. It developed 400 hp and was relatively light despite its iron blocks, thanks to using high speed instead of large displacement to develop its power. Only a single prototype was built in France, but the American purchasing mission picked up rights to it, and a few were made by Duesenberg at Elizabeth, New Jersey. The American version was somewhat redesigned by a Colonel King, who removed some of the characteristic Bugatti peculiarities,[6] Apparently no example of the Bugatti-King ever flew, although it performed well on the test bench. In any case, it is hard to see how any reasonable person could have seriously considered putting any Bugatti design into volume production (Fig. 5-14).

The other French engine of automotive derivation was the Delage. The remarkable thing about this engine was that it was designed as an airplane's nose. That is, the outside front part of the engine was very well streamlined and needed only a minimum of partial cowling (Fig. 5-15). Actually there were two Delages, one 488 cu in. (8 L) and one 732 cu in. (12 L), both inverted V-12s. The 488-cu-in. engine was intended for a Coupe Deutsch racer. It was rated for 400 hp at 3800 rpm, geared to 1900 at the propeller, and, as

[6]Ettore Bugatti was especially bitter about the use of rational intake and exhaust port shapes.

FIG. 5-14 Bugatti-King H-16, 400 hp, 1918. Bugatti cognoscenti will note numerous departures from Bugatti automobile practice—the numerous compound-curved surfaces and the accessibility of the cylinder hold down studs especially. Unlike the elimination of dead pockets of cooling water in the passages and the curved intake and exhaust ports, these were not Colonel King's modifications; they were largely present in Ettore Bugatti's original design. *(Courtesy of NASM-Smithsonian.)*

FIG. 5-15 Delage V-12. The streamlined shape is not a cowling, but is the shape of the actual engine. *(Courtesy of the National Archives.)*

the 2 to 1 reduction suggests, used the timing gears for reduction gearing, as did the early Renaults. The Delages were among the very few aircraft engines to use a Roots compressor for a supercharger; centrifugal blowers absorb less power. The Delage firm's bankruptcy in 1933 prevented these interesting concepts from being further developed.

Germany—The D-B 600 Series and Jumos

The Germans liked to blame the restrictions of the Treaty of Versailles for what they claimed to be a lag in their engine progress. The facts are otherwise. They violated the treaty by building large engines but persisted of their own volition in staying with separate-cylinder engines as late as 1936. But when they did build cast-block engines, they built first-class ones.

There is a gray area of history in the German efforts to redirect their aircraft engine development. Beginning in 1934, the Nazi government paid a subsidy to Daimler-Benz and Auto-Union (DKW, Horch, Wanderer, and Audi) in support of their international auto racing costs. The resulting cars dominated European racing, except for some late victories by the Italian Alfettas, right up to the start of the war. It has been claimed that the real reason for the subsidy was to underwrite aircraft engine development and that the engines that powered Messerschmitts in the war were outgrowths of the subsidized Grand Prix racers. Mercedes has said, in refutation, that the subsidies did not fully cover the firm's racing costs and had nothing to do with aircraft engine development.

The peculiar nature of the racing formula in effect at the time contributed to the controversy. Most racing formulas have been based on engine displacement. Under such rules, small, very high speed engines, having no relation at all to aircraft needs, become the norm. The thirties formula, though, was simply a limitation of maximum weight to 1650 lb (750 kg). Under this formula, a large, light, relatively slow-running engine—in other words, something very like an aircraft powerplant—could be used if the designers believed that better performance could be had that way.

Two very different engines resulted from the German subsidy. The Mercedes was a high-revving straight eight in the classic racing engine tradition, but the Porsche-designed Auto-Unions were more like aircraft engines. These were, first, a sixteen and, later, a twelve, of large size rather than high speed. One interesting feature of the Auto-Unions was the use of three camshafts. Each bank of cylinders had its own camshaft to operate the exhaust valves, and a central camshaft operated the intake valves for both banks. The Auto-Union might have had some potential for aircraft use, but if so, it was never explored.

The probable truth about the subsidies is that the Nazis hoped for both technical and propaganda benefits from the racing program. They got neither; indeed, the overall effect was probably disadvantageous. The racing victories were widely discounted as being due more to big money than to talent, and the resources put into race car engines might have been better employed elsewhere.

Whether or not this diversion of resources had anything to do with it, the first German cast-block engines did not appear until 1936. There were two of these, the Mercedes D-B and the Junkers Jumo. Both were inverted V-12s with simple, sturdy construction and large displacement in relation to their horsepower.

The Mercedes, usually referred to as simply D-B, was one of those rare designs that are right from the start. All the subsequent wartime D-B engines were directly derived from it. It displaced 2069 cu in. (33.9 L), practically as much as the 12Y Hissos, but was considerably less bulky; such D-B-powered planes as the Me 109 show far better entry lines than any Hisso-powered fighter. Aside from the peculiarity of being inverted, the D-B 600 had three unusual features. It had roller-bearing connecting rods, which are more indicative of distrust in one's ability to make good plain bearings than of superior engineering;[7] it used dry cylinder liners,[8] a retrograde feature that may have stemmed from fear of porosity in the block castings; and it

---

[7]Later D-B models did in fact go to plain bearings.
[8]The poor transfer of heat to the coolant, resulting from the use of dry sleeves, was probably the reason that the D-B 600 series needed larger oil coolers than comparable engines. More of their heat dissipation took place through the oil, and less through the glycol, than was the case for the Merlin and the Allison.

employed a unique method of holding the blocks to the crankcase. Instead of using studs, the D-B had threaded inner ends on the cylinder sleeves. When the block was placed on the crankcase, thin-walled serrated nuts were threaded onto these sleeves and tightened against the inner surface of the crankcase. This was an excellent feature, saving the weight of the studs and avoiding any possibility of distortion.

Within a year or so the D-B 601 replaced the 600. It differed from its predecessor in only two important ways: the use of direct fuel injection and the fluid-clutch supercharger drive. The fluid coupling was so connected to the outside atmosphere as to slip more when the ambient atmospheric pressure was high. This very clever device constituted an automatic altitude compensator, making the blower run faster at higher altitudes. Horsepower went up from the 600's 950 hp to 1050, and fuel consumption was reduced.

While various Nazi warplanes used engines in the D-B 600 series, the classic application was that in the Me 109. The development of the engines closely parallels the evolution of this most-produced of all fighters. Actually the little-known A, B, and C models of the 109 had Jumo power, but it was with the D-B 600 –powered D model that, in Spain,

the Me 109 came into its own. The E model of late 1938 used the D-B 601A, and the F, possibly the best of all 109s, used the 601E. The 601E had a larger supercharger and could deliver up to 1300 hp for what Americans called METO (Maximum Except at Takeoff) power.

In 1941 the 603 began to supplement the 601. This was an even larger engine at 2715 cu in. (44.5 L). The need to go to such extremes of size to get the needed power, in this case 1750 hp, may be considered as evidence of inferior engineering in comparison with what Rolls-Royce and Allison were able to do. It is true that all the D-Bs (except the 610) were trouble-free powerplants in the field, but so were the Allisons. So much additional bulk and weight does not seem to have been justified (Fig. 5-16).

Where the 600 and the 601 had used side-by-side connecting rods, like a modern V-8 automobile engine, the 603 used the older system of a master rod and a link rod for each pair of cylinders, like the 12Y Hispano. For some reason, possibly connected with this system, the 603 had different compression ratios for each bank; the left side had 7.5 to 1, the right 7.3 to 1. Ultimately some models of the 603 were made to give 2830 hp at 3000 rpm, a rather startling speed for such a big engine.

FIG. 5-16 Mercedes D-B 600 series engine in Messerschmitt 109. (*Courtesy of NASM-Smithsonian.*)

The D-B 603's weight of 2002 lb (910 kg) interdicted its use in German single-engine fighters. Allied and Japanese fighters, having wingspans of from 36 to 40 ft (11 to 12 m), could carry the big engines necessary for combining range and performance,[9] while the 32-ft 6-in. to 34-ft (9.9- to 10.36-m) span German fighters, good as they were, had the fatal defect of not being able to carry the fight to the enemy. The 603's service was confined to larger aircraft such as the twin-engine Me 410, the Do 271M bomber, the He 219 night fighter, and the remarkable experimental Do 335 push-pull twin.

The next Daimler-Benz development was the 605. This was a 601 enlarged to 2179 cu in. (35.7 L) and intended for operation at higher rpm. The A, B, and C models gave 1475 hp at 2800 rpm and went into the G series Me 109. The extra power produced little in the way of improved performance because of the extra weight of both the new engine and the additional armament loaded onto this small airframe. The 605 series went to a −L model, giving 1700 hp with a two-stage supercharger, and an AM with water-methanol injection of METO power and 1800 hp.

Water-methanol injection is a way of permitting short bursts of exceptionally high power; by spraying a little water into the mixture, the combustion temperature is brought down below dangerous levels. The alcohol is not the essential element, its function being largely that of an antifreeze for the water in the tank. It seems to have been an American invention originally, but all the combatants in the Second World War used it.

Daimler-Benz had a great deal to do with one of the Luftwaffe's worst fiascos, the He 177. Opinion varies as to whether this was a two- or a four-engine craft, since its two propellers were each driven by a pair of D-B 605s. Whether Daimler-Benz's design or Heinkel's installation was at fault is hard to determine, but the combination, which was called a D-B 610, was extraordinarily fire-prone. Heinkel himself attributed the trouble to the

placement of oil lines too close to the exhaust pipes, but whatever the reason, what could have been a formidable 5900-hp bomber was a total loss and a great waste of a small work force and scarce resources.

The Jumo (JUnkers MOtoren) was similar in concept to the Mercedes in that it was an inverted twelve and was large for its power. Structurally it differed from the D-B in that it used cylinder blocks cast in one piece with the crankcase. This strong and neat construction was made possible by splitting the block on its vertical centerline. The first cast-block Junkers engines were the 1202-cu-in. (19.7-L) 210 and the 2136-cu-in. (35-L) 211, both brought out in 1936. By 1941 the 210 was giving 700 hp when equipped with a two-stage supercharger, a respectable figure for its size, and the 211 was giving 1100 hp—not much for its size but good for its weight of 1356 lb (616 kg). The 210 was not a particularly useful size and was apparently dropped in 1942, but the 211 was brought up to 1400 hp in the J model (Fig. 5-17).

Although Jumos saw service in the first Me 109s, in general they were bomber engines. Their excellent specific consumption, 0.46 lb (0.209 kg), made them especially desirable for bombers such as the Ju 88, the He 111H, and the Focke-Wulf Ta 154. The principal single-engine application for Jumos was in the Ju 87, the well-known Stuka. Highest Jumo power was 1776, in the 213 E of 1942. The Jumo was a workhorse rather than a powerhouse, a fact that may be related to Junkers's background in aircraft Diesels.

The Allison—and Two Hopefuls

The only American liquid-cooled engine to see service in World War II was the Allison V-1710. Allison was an Indianapolis firm that had done well in a small way with Liberty modifications and with reduction gears for others' engines. The Liberty modifications, described elsewhere, were done on Army Air Corps contracts and included a certain amount of quantity production of the inverted model.

Around the time that the Army was washing its hands of the Curtiss Conqueror, Allison began

[9]Of course, the fact that the Merlin in the P-51 gave the power without being big and heavy was largely what made it the most effective of all.

FIG. 5-17 Junkers Jumo 210 inverted V-12, 1940. Construction and appearance of the larger 211 was almost identical. *(Courtesy of the National Archives.)*

to develop its own engine. The target was 1000 hp, and Allison intended that the engine should be large enough to deliver this power easily. Development proceeded slowly until the Navy entered the picture. The Navy, while not losing its attachment to air-cooled power for airplanes, needed liquid cooling for dirigibles and was not altogether happy with its current dependence on Maybach. It somehow persuaded itself that the Allison might make a good dirigible powerplant and awarded a development contract to the Indianapolis firm. When this particular piece of nonsense petered out with the loss of the Macon in 1935, James Allison sold out to Eddie Rickenbacker, who immediately resold it, at a good profit, to General Motors. The firm today remains a GM component under the style of "Detroit Diesel Allison Division." GM, with the unfailing acuity that it showed throughout the long reign of Alfred P. Sloane, enlarged the plant and provided the capital to get the V-1710 to the production stage.

An interesting incident of this period in the Allison's evolution will illustrate the fact that intuition still had a place in engine design. Fred Duesenberg, whose great cars were built nearby in Indianapolis, visited the Allison plant a couple of years before his death and was shown one of the early engines. Duesenberg liked the design except for a gut feeling that the crankshaft was a little light. The Allison engineers assured him that the design had been carefully calculated and that it had ample strength—but the major finding from the testing was that the crankshaft needed beefing up.

The production Allison turned out to be the sturdy and reliable powerplant that its designers had striven for. The only thing that stood between the Allison and real greatness was its inability to deliver its power at sufficiently high altitudes. This was not the fault of its builders. It resulted from an early Army decision to rely on turbosupercharging to obtain adequate power at combat heights. Even this decision was not a technical error; a turbosupercharged Allison was as good a high-altitude engine as most. The trouble was that the wartime shortage of alloying materials, especially tungsten, made it impossible to make turbosuperchargers for any but a small proportion of Allisons. Bomber engines got the priority (Fig. 5-18).

The few turbosupercharged Allisons that were made were allocated to P-38s, making the high-altitude performance of that plane its best feature. All 14,000 P-40s got gear-driven super-

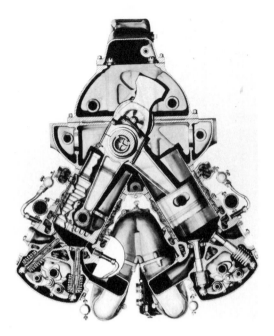

FIG. 5-18(a)   Cross section of Allison V-1710 engine. Compare with Merlin section, Fig. 5-10(b). The Allison's "lower end" looks a bit sturdier but the main design difference is at the head. The Allison used splayed valves but not, as in many separate-cylinder engines and most radials, to permit the use of two large valves per cylinder—it employed four. The splayed valves gave the Allison a "pent-roof" combustion chamber, very much like that of the Offenhauser-Meyer-Drake engine that dominated Indianapolis racing for so long. Many features of cast-block construction show up clearly here. Note also the hollow sodium-filled valves.

chargers and, as a result, were never first-class fighter planes. Donaldson R. Berlin, the P-40's designer, has said that P-40s experimentally equipped with turbosuperchargers outperformed Spitfires and Messerschmitts and that if it had been given the engine it was designed for, the P-40 would have been the greatest fighter of its era. This may be to some extent the bias of a proud parent, but there is no doubt that the deletion of the turbosupercharger ruined the P-39. This fighter, with the high span loading that resulted from its small span of 34 ft (10.36 m), as against the P-40's 37 ft (11.27 m), was only useful as a ground-attack plane in the absence of the turbosupercharger it

was designed to use. Bell tried to remedy this deficiency in the P-63 Kingcobra by increasing the span to 38 ft (11.58 m), but even so, the P-63's service ceiling was 43,000 ft (13,106 m), and its best speed came at 24,450 ft (7452 m), as against 46,000 and 30,000 ft (14,020 and 9144 m) for the service ceiling and height for best speed, respectively, of the original turbosupercharged P-39.

Had Allison's engineers been able to put the effort into gear-driven superchargers that Pratt and Whitney and Rolls-Royce did, it might have been a different story. As it was, there can be little doubt that the V-1710 had more potential than was actually exploited. The extent of this potential can be seen in the fact that the Allison has a definite following and may even be preferred to the Merlin in present-day Unlimited Class air racing. Its slightly greater displacement, good heat dissipation, and rugged lower end allow it to be successfully "souped up" to higher power than the Rolls. Allison-powered P-51s are, often as not, victorious over Merlin-powered versions of the same plane. The Allison served its country well, but had it been given the chance, it could have contributed much more to victory than it actually did.

In addition, there were two experimental liquid-cooled engines of considerable promise, neither of which was developed into a production model. Best known of these was the Lycoming O-1230, developed on an Army Air Corps contract of around 1938. This was one of a number of contracts intended to develop what was called a "Hyper" engine, having considerably more specific output than any existing engine. When Lycoming's opposed-twelve Hyper was announced in May 1940, it gave 1200 hp from 1230 cu in. (20 L), a specific output of 1 hp/1.025 cu in. (60 hp/L). It was an attractively neat engine, very slim when seen from the side. It was intended for buried installation in the wings of large planes, but, although one occasionally heard news of it, it was never developed further.

Obviously it was the wrong size; 1200 hp wasn't enough. Had it been even 25 percent larger, it would have been a useful bomber engine, giving significant improvements in range through the drag reduction inherent in its compact flat shape. Of course, both the B-17 and the B-24 used 1200-hp

FIG. 5-18(b)   Allison V-1710 engine, about 1250 hp, 1938–1946. Probably for the P-39—the front end looks as if it is intended to be connected to a driveshaft rather than to a propeller. (Courtesy of Detroit Diesel Allison Division, General Motors Corporation.)

engines, but there was no way the O-1230 could have been substituted for the standard radials. To use buried engines, the whole wing structure would have needed redesigning and the weight and balance rearranging. Still, let your imagination play around with the concept of a midget B-36, somewhere around B-24 size, with four of those flat engines buried in the wing!

One wonders about the persistence of official interest, during the thirties, in liquid-cooled engines in the 1200-cu-in. (21-L) class. The British kept prodding at the old Kestrel design, the French had the 12X Hissos and the Farman 18T, the Germans built the Jumo 210, and America sponsored the Lycoming and others. To some extent, the American interest can be justified by the Hyper concept, since it offered the hope that a 1200-cu-in. (21-L) engine could do the same job as existing larger and heavier motors. Otherwise, it would seem to have been rather obvious that nothing under 1650 cu in. (27 L) would be powerful enough for adequate performance in fighters and the faster types of bombers. There was exactly one World War II combat plane that used such engines, the Westland Whirlwind. This was a sort of "light twin" fighter,

a most beautiful airplane, by the way, and it might have been a *succès d'estime* had not the heavy demands made on Rolls by other work forced it to abandon development of its Peregrine[10] engine.

The other abortive American liquid-cooled engine was the Ford XV-1650, an experimental V-12 of 1941. This was a well-engineered design, incorporating several features that were expected to raise efficiency and reduce manufacturing labor-hours. It used steel castings instead of the usual forgings, Ford then being well in advance of the rest of the world in this economical technique, and side-by-side connecting rods like an automotive V-8. Both a turbosupercharger and a fuel injection system were integral parts of the design. By blowing excess air through the cylinders during the valve lap period (see Chap. 1), cooling was aided to the point where an output of 2000 hp was considered attainable. The specific output would have been an astounding 1 hp/0.825 cu in. (74 hp/L), well into the Hyper category. Ford built and tested a 2-cylinder model to confirm these figures, but the

[10]The Peregrine was a "Merlinized" 885-hp outgrowth of the Kestrel.

design was never proceeded with. Henry Ford was in bad odor with the administration, and this may have influenced the government's decision; but the fact was that 1941 was rather late in the game. It would probably not have been ready in time. The war was fought, generally, with engines that had been well along by 1940.

The Last of the Liquid-Cooled Engines

The story of the Rolls-Royce Eagle II supports this supposition. The second Eagle had its origins in rivalry between Rolls and Napier. Rolls, right at the beginning of the war, undertook the design of an engine that would fit into any aircraft that used the Merlin but which would have considerably more displacement and output. Using the same bore and stroke as its R-type Schneider Cup engine of 10 years before, it built the 2240-cu-in. (36.7-L) Griffon to fit this specification. The Griffon came out at 1750 hp, got fuel injection on the −72 model built for the Fleet Air Arm, and ended with 2420 hp in the 101 series using a three-stage supercharger (Fig. 5-19).

The Griffon did a fine job, notably on late marks of the Spitfire, and, 30-odd years after the war, took the world's absolute speed record away from Daryl Greenamyer's Bearcat. Nevertheless, it was challenged and even somewhat eclipsed by the Napier Sabre. The Sabre resulted from combining Napier's long-standing attachment to the H-type configuration with the lessons that Bristol had taught anent the virtues of sleeve valves. It took the form of two opposed twelves atop one another and had exactly the same displacement as the

FIG. 5-19 Rolls-Royce Griffon, probably an early single-stage-supercharged type of about 1730 hp. Most notable differences from the Merlin are the bulges at the front of the camshaft covers and the front mounting of the magneto, both resulting from taking the timing drive from the front of the crankshaft. This was done to reduce torsional vibration of the crankshaft. *(Courtesy of Rolls-Royce Limited.)*

Griffon. It was considerably more powerful than the Rolls engine, delivering 2200 hp at 3700 rpm at first, 2400 hp at the end of the war, and eventually 3000 hp. Better at low altitudes than at high ones, it made its mark as the powerplant for the Hawker Typhoon, Britain's counterpart of the P-47 as a fighter-bomber. At first it justified the apprehensions natural to the introduction of a complex 24-cylinder engine by being extremely unreliable in service. However, this turned out to be mainly the fault of the squadron mechanics who (knowing more than the engineers did, as mechanics so often do) had opened and misadjusted the sealed automatic altitude-compensating control boxes (Fig. 5-20).

However, if Napier had the power, Rolls had the connections in high places and was not averse to exploiting the Sabre's early troubles. Accordingly, it was allowed to proceed with the Eagle as its answer to the Sabre despite the lateness of the hour and the near-certainty that it would emerge too late to take part in the war. The Eagle II, which was so large that I am reserving a description of it for the last chapter, did not, in fact, arrive on the scene in time for combat use. Such engines having no place in the peacetime scheme of things, the Eagle II did really represent a grievous waste of resources.

Since the Merlins and Sabres stayed in service until the advent of the turbines, the story should logically end here. But odd things happen. After the war, Italy experienced a resurgence of national vigor that, at times, took some improbable directions. One of these was Fiat's development of its A-38 R/C 15-45, an inverted V-16 of 2117 cu in. (34.7 L). It probably never flew. It gave to Italy the strange distinction of being the last to build both a separate-cylinder and a cast-block liquid-cooled engine.

Russian Inlines: Almost an Afterword

The story of Russian liquid-cooled development is sketchy and, as so often with Soviet technical history, full of unanswered questions. Their engine designations come from the name of the "design bureau" head, and since a given design bureau may have been associated with more than one type of engine, the designation is less than revealing. For instance, the AM line takes its name from Mikuline, but there were AMs other than liquid-cooled inlines; and the M series is also supposed to be named for Mikuline.

The Ms were thought, in the West, to be derived from the Curtiss Conqueror. Pre-war figures of close to 1000 hp have been cited, but they

FIG. 5-20   Napier Sabre, 2300 hp, 1941. (*Courtesy of the National Museum of Science and Technology, Ottawa, Canada.*)

were not prominent during the actual combat. The AM-30 line is said by some Western sources to have been based on the 12Y Hispano, but Soviet releases have said that the AMs derived their basic design from the BMW VI and that it was the VK series that was based on the Hisso. If the AM engines did in fact evolve from the BMW, they represent a unique case of a welded-cylinder engine being turned into one with a cast block.

Be that as it may, the AMs were big engines, at 2850 cu in. (46.7 L). The best known of these, the AM-38, gave 2150 hp from a relatively light 1892 lb (860 kg). The similar AM-42 gave a bit less power: 2000 hp. One circumstance makes it probable that these are accurate figures: the fuel-consumption figures given in the same release are not boastful. Specific consumption is quoted at a high 0.625 lb (0.284 kg) at full power and as 0.594 lb (0.27 kg) at 1410-hp cruise power.

The VK figures hint at an interesting development history. The VK-105 displaced 2147 cu in. (35.2 L) and delivered 1100 hp, minimal for wartime service but better than any production-model French Hispano at the time (1940). The second version, the VK-107, had reinforced cylinders (!) that must have resulted in a reduction in bore, because displacement dropped to 2075 cu in. (34 L), but horsepower rose to 1600. Photos of these engines show the usual rough finish and inelegant design that characterized Soviet aviation up to the time when jet speeds made roughly built aircraft unworkable, but, as with so many Russian products, they worked well enough to do the job.

That job was not the same as those assigned to the aircraft of the other fighting nations. The primary task of the Red Air Force was close-in tactical warfare, as exemplified by the tank-busting IL-2 and IL-3. Furthermore, with its cold appreciation of the difference between efficiency and effectiveness, the Soviet military accepted high attrition rates as long as the job got done (elsewhere, such high attrition rates would have forced a reappraisal of tactics). Thus, the Russian engines had to be suitable for mass production and capable of very high output at low altitudes but could sacrifice fuel economy, performance at high altitudes, and TBO. Soviet engines, like Soviet fliers, were effective but expendable.

Data for Engines

Maker's name	Date	Cylinders	Config- uration	Horse- power	Revolutions per minute	Bore and stroke, in. (mm)	Displace- ment, cu in. (L)	Weight, lb (kg)	Remarks
Hispano-Suiza	16	8	V, L	180	1600	4.7 (120) × 5.1 (130)	718 (11.8)	467 (212)	Geared version developed 220 hp
Hispano-Suiza model 42 (8 F6)	18	8	V, L	300	1800	5.5 (140) × 5.9 (150)	1127 (18.4)	528 (240)	Called "model H" when made by Wright
Curtiss K-12	17	12	V, L	400	2500	4.5 (114) × 6 (152)	1145 (18.8)	679 (309)	
Curtiss K-6 (and C-6)	18	6	I, L	150	1700	4.5 (114) × 6 (152)	573 (9.4)	417 (190)	
Curtiss D-12	22	12	V, L	350	1800	4.5 (114) × 6 (152)	1145 (18.8)	704 (320)	Service versions gave 400 hp
BHP	17	6	I, L	230	1400	5.7 (145) × 7.5 (190)	1148 (18.8)	680 (309)	
Rolls-Royce Kestrel	27	12	V, L	490	2250	5 (127) × 5.5 (140)	1296 (21.2)	865 (393)	Service models gave up to 690 hp; 1939 Peregrine had Merlin design features, Kestrel size, gave 885 hp, but was troublesome
Curtiss Conqueror	28	12	V, L	600	2400	5.1 (130) × 6.2 (159)	1570 (25.7)	770 (350)	As high as 800 hp geared and supercharged
Rolls-Royce R	29	12	V, L	2500	3200	6 (152) × 6.6 (170)	2240 (37)	1630 (740)	Developed from 955-hp Buzzard of 1928; Kestrel design features
Rolls-Royce Merlin I	36	12	V, L	1030	3000	5.4 (137) × 6 (152)	1650 (27)	1320 (600)	Original (1934) Merlin gave 790 at 2500 rpm; 1946 Merlin 130 had fuel injection, gave 2030 hp
Hispano-Suiza 6Mb	28	6	I, L	300	2100	5.1 (130) × 6.7 (170)	831 (13.6)	550 (250)	
Hispano-Suiza 12Ycrs	33	12	V, L	860	2400	5.9 (150) × 6.7 (170)	2196 (36)	1014 (461)	World War II fighter engine
Hispano-Suiza 12Z-11Y	47	12	V, L	1500	2600	5.9 (150) × 6.7 (170)	2196 (36)	1364 (620)	
Farman 12 WE	24	12	W, L	500	2150	5.1 (130) × 6.3 (160)	1555 (25.5)	1034 (470)	

(cont.)

Data for Engines (continued)

Maker's name	Date	Cylinders	Configuration	Horsepower	Revolutions per minute	Bore and stroke, in. (mm)	Displacement, cu in. (L)	Weight, lb (kg)	Remarks
Farman 8 VI	21	8	Iv, V, L	350	2500	5.3 (135) × 5.5 (140)	977 (16)	704 (320)	
Farman 18 W.I.	26	18	Iv, W, L	730	3400	4.3 (110) × 4.9 (125)	1304 (21.4)	701 (319)	
Farman 12 W.I.	32	12	Iv, V, L	550	2260	5.3 (135) × 5.5 (140)	1466 (24)	902 (410)	
Farman 12 Brs	32	12	Iv, V, L	460	4020	3.6 (90.5) × 3.9 (100)	471 (8)	565 (257)	
Farman 18T	30	18	Iv, T, L	1200	3400	4.7 (120) × 4.7 (120)	1498 (24.5)	1062 (483)	
Lorraine Courlis	29	12	W, L	600	2000	5.7 (145) × 6.3 (160)	1960 (32.1)	935 (425)	
Lorraine Petrel	32	12	V, L	500	2250	5.7 (145) × 5.7 (145)	1803 (29.5)	1045 (475)	
Lorraine Eider	28	12	V, L	1050	2200	6.7 (170) × 6.5 (165)	2743 (45)	1397 (635)	
Bugatti (Bugatti-King)	18	16	H, L	410	2100	4.3 (110) × 6.3 (160)	1484 (24.3)	1248 (567)	
Delage C.I.D.R.S.	31	12	Iv, V, L	400	3800	3.9 (100) × 3.3 (84.4)	488 (8)	814 (370)	Used Roots (positive displacement) blower as supercharger; centrifugal type is usual
Delage C.I.D.R.S.	31	12	Iv, V, L	450	3600	4.3 (110) × 4.1 (105)	732 (12)	968 (440)	
Mercedes-Benz D-B 600	38	12	Iv, V, L	1000	2400	5.9 (150) × 6.3 (160)	2069 (33.9)	1510 (686)	
Mercedes-Benz D-B 601	38	12	Iv, V, L	1360	2600	5.9 (150) × 6.3 (160)	2069 (33.9)	1540 (700)	To 1700 hp as early as 1940
Mercedes-Benz D-B 603	41	12	Iv, V, L	2830	3000	6.4 (162) × 7.1 (180)	2715 (44.5)	2002 (910)	
Mercedes-Benz D-B 605	42	12	Iv, V, L	1475	2800	6.1 (154) × 6.3 (160)	2179 (35.7)	1663 (756)	
Mercedes-Benz D-B 610	40	24	Iv, W, L	2450	2800	6.1 (154) × 6.3 (160)	5438 (71.4)	3476 (1580)	

Engine		Cyl	Type			Bore × Stroke	Displacement	Weight	Remarks
Junkers Jumo 210	36	12	Iv, V, L	700	2600	4.9 (124) × 5.4 (136)	1202 (19.7)	968 (440)	Weight is for G model
Junkers Jumo 211	36	12	Iv, V, L	1100	2300	5.9 (150) × 6.5 (165)	2136 (35)	1356 (616)	1400 hp maximum, in J
Junkers Jumo 213	42	12	Iv, V, L	1776	3250	5.9 (150) × 6.5 (165)	2136 (35)	2024 (920)	
Allison V-1710-G6	41	12	V, L	1250	3200	5.5 (140) × 6 (152)	1710 (28)	1595 (707)	Typical wartime model
Lycoming 0-1230	40	12	O, L	1200	3100	5.3 (133) × 4.8 (121)	1234 (20)	1325 (602)	
Ford XV-1650	41	12	I, V, L	2000	NA	NA	1650 (27)	NA	Projected but not built
Rolls-Royce Griffon II	40	12	V, L	1735	NA	6 (152) × 6.6 (170)	2240 (37)	2090 (948)	Griffon 67 gave 2375 hp
Napier Sabre	40	24	H, L	2200	3700	5 (127) × 4.8 (120)	2240 (37)	2500 (1134)	Type-tested at these figures; production engines more powerful
Mikulin AM-38	NA	12	V, L	1600	2150	6.3 (160) × 7.5 (190)	2850 (46.7)	1892 (860)	Late versions gave 2150 hp at low altitudes
VK-105	NA	12	V, L	1100	2700	5.9 (148) × 6.8 (170)	2147 (35.2)	1270 (575)	
VK-107	NA	12	V, L	1600	NA	5.7 (145) × 6.8 (170)	2075 (34)	1655 (748)	Bore calculated from displacement; published figure is 5.9 (148), which is wrong

I—inline; V—V-type; L—liquid cooled; W—W-type; T—T-type; H—H-type; Iv—inverted; O—horizontal opposed; NA—not available. Open figures—U.S. Customary; figures in parentheses—metric. Weights are dry.

SIX

The radial engine to 1930.

Rationale of the Radial Engine

The stationary radial looks and works like the rotaries described in Chap. 4, the difference being that the engine stands still and the crankshaft turns. This configuration offers many advantages over the inline type, especially when it is used in connection with air cooling. The short crankshaft gives compactness in the fore-and-aft dimension and rigidity without weight. It is well suited for air cooling, since each cylinder is impinged on by its own unobstructed flow of cool air. Because all cylinders act on the same crank throw, a major source of vibration in inline engines, called "rocking couple," does not exist. Accessibility for servicing is excellent, and the absence of a radiator minimizes the susceptibility to battle damage.

Before these advantages could be exploited a number of problems had to be solved. How were the cylinders to be connected to the single crank? What was to be done about an oil sump when the bottom of the crankcase had to be open to 1 or more cylinders? How should cylinder heads be designed to cool adequately in an airstream moving only as fast as the aircraft and not speeded up as on a rotary? How was even distribution of fuel mixture to the cylinders to be achieved? How was the drag inherent in a large-diameter engine to be minimized?

Manly's 1902 radial had embodied solutions to many of these problems, but his solutions, impeccable though they were in terms of his time, became less and less acceptable as time passed and technology advanced. The use of slipper bearings to connect the slave rods to the master rod was a dead end, though it died hard. Manly's total-loss oiling system, never desirable, became less and less acceptable as longer and longer flights became the norm—its use in rotaries was one of the reasons for the demise of the type. Gasoline at the turn of the century was extremely volatile, and so the simple intake manifold of the Manly engine sufficed to assure good mixture distribution, but something better was needed for the gasoline of 20 years later.

This matter of even mixture distribution may need explaining. If the carburetor did a perfect job of vaporizing gasoline and mixing it with air, the intake manifold would only have to be a well-designed arrangement of ducting to deliver the same amount of the same mixture to each cylinder. Unfortunately, all carburetors are rather imperfect devices, and so the manifolding arrangements are required to make up for some of the carburetor's deficiencies. These deficiencies were well demonstrated by a classic investigation performed by Sun Oil in 1946. On a conventional 6-cylinder car engine, mixture distribution at part throttle was found to be appallingly bad. Each cylinder got a different total amount of mixture, a different air-to-fuel ratio, and even a mixture of a different octane rating.

Some small radials in the late twenties solved the distribution problem by heating the mixture, but this robbed too much power to be acceptable for commercial or military use. Until fuel injection became available, the best answer turned out to be at least a mild degree of supercharging. This improved mixture distribution in two ways. First, the blower thoroughly agitated the mixture, completing the mixing job begun by the carburetor. Second, the supercharger slightly pressurized the whole manifold, guaranteeing equal pressure at each intake port during the time the intake valve was open.

The solution to the problem of having many cylinders on a single crank lay in using link rods, articulated to a master rod, as pioneered by Gnome (Fig. 6-1). The geometry of this arrangement is imperfect, and many 5- and 3-cylinder engines using it have been decidedly rough running, but it works adequately well on multicylinder engines. It eventually achieved universal acceptance.

Also available early as a solution to the lubrication problem was the dry-sump system, as described in connection with the Austro-Daimler in Chap. 3. In most radials a small additional refinement was added. The inner ends of the cylinders were allowed to extend some distance into the crankcase. Thus, surplus oil gathered around the outside of the lowest cylinders instead of flowing down into them. Even this is imperfect. When an engine has stood for some time, it has occasionally

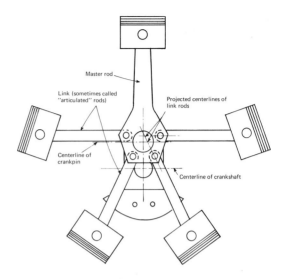

FIG. 6-1 Piston and connecting rod assemblies of a 5-cylinder radial engine shown in position on partly assembled crankshaft. Note that projected centerlines of link rods do not coincide with centerline of crankpin. This offset shifts continually as the engine runs, and the link rods line up with the crankpin centerline twice in each revolution.

happened that enough oil has drained down past the piston rings to cause damage from hydraulic lock when the engine was started. It is good practice to turn a radial or even an opposed engine over by hand before starting it to assure that there is no accumulation of oil in the cylinders.

The drag problem was solved when it needed to be solved. Until sometime in the mid-twenties, anything over 100 mph (161 km/hr) was considered a pretty decent cruising speed, and at such speeds, the drag of an uncowled radial is of little importance. But by the later part of the decade, speeds of 150 mph (242 km/hr) were becoming common, and the drag of a given component is 2.25 times as great at 150 mph as it is at 100. Had the Townend Ring and the NACA cowl (see Chap. 1) not come along when they did, it is probable that the drag of exposed radial engine installations would have caused the radial to be relegated to a secondary position.

It took years of experimentation before the problem of cooling large high-output cylinders by an impinging airstream was solved. There were successful small air-cooled radials—the Anzanis—by 1910 and a good large water-cooled one as early as 1913, but it was not until late 1918 that the first good large air-cooled radial was built. A great deal had to be learned, and unlearned, first.

French and British Developments, 1913-1919

The first successful large radial was the Salmson Canton-Unne of 1913. Except for being water-cooled, the Salmson demonstrated that all the theoretical advantages of the radial concept were in fact real. It opened the way for all future large radial engines. Salmson was a firm that is hard to describe in present-day terms to an American, though Britons will comprehend it very well as a "general engineering" company. Branching out from its original pump business, it made very good engines and then developed an airplane to use them. It got out of the airframe business after World War I and built some good small- and medium-size air-cooled radials (described in Chap. 8). It made automobiles ranging from an exquisite 45-cu-in. (750cc) racing car to an indestructible small family car of no performance whatever and, sad to say, ended up where it had begun—in the pump business.

The first Salmson had 7 cylinders and gave 90 hp at 1250 rpm. Almost simultaneously a nine, giving 110 hp at 1300 rpm, was introduced. The Canton-Unne part of the name came from the patent owners of the connecting rod system that Salmsons used. There was no master rod; all 9 cylinders used link rods. These link rods went to a cage that rode on the crankpin and was connected by gears so as to always remain vertical while the shaft turned; it revolved, or orbited, but did not rotate. Salmson evolution was continuous during the war years, culminating in the Z-9, usually described as a 270-hp engine. Variously redesignated as the R-9A and the AC9, it was available as late as 1924, but it was obsolete by 1920. The Z-9 normally came complete with radiator and cowl, making a compact "power egg." It was equipped with ball bearings for both the mains and the connecting rod cage and was durable without being

heavy. It was also efficient, having a specific consumption of 0.495 lb (0.225 kg). The Salmson power egg included radiator flaps for good temperature control; no doubt this made its contribution to fuel economy.

Comparison between Salmson's first and last water-cooled radials typifies the progress in engine design that marked the 5 frantic years that lay between them. Specific output rose from 7.9 to 4.8 cu in./hp (from 7.7 to 12.2 hp/L). This was an increase of 61 percent in specific output, accomplished while rpm rose only 20 percent. The Salmson's success inspired one imitation, the Fiat A-18 of 1920. Like so many Italian engines, the A-18 was both a fine-looking engine and, at 320 hp, a powerful one, but by that time it was clear that the future of the radial lay with air cooling, and the Fiat was dropped after only a year or two.

Possibly Salmson was prompted to get into the airframe business by a deplorable flying machine called the Salmson-Moineau. The Salmson-Moineau was a biplane with a Canton-Unne engine set sidewise within the fuselage, with driveshafts extending out between the wings to drive two propellers through bevel gears. It has been said that no airplane with a driveshaft was ever any good, and the Salmson-Moineau probably contributed to this belief. For the most part, driveshafts and remote gearsets require more rigidity than can be afforded within the weight limitations of an aircraft.

In contrast, Salmson's own aircraft was by all accounts a first-class piece of work. Called the S.A.L. 2-A.2, it was a two-place observation and light bombing biplane, rather like the Renault and the Breguet with its equal-span, slightly back-staggered wings (Fig. 6-2). The Americans used them for a while and were not delighted when they had to give them up on the arrival of the de Havillands. While they went less than 100 mph (161 km/hr), compared with the 120 mph (200 km/hr) claimed for the de Havillands, their engines always kept running, they handled well, and they were hard to flame—virtues not shared by the D.H. 4.

The first successful large air-cooled radials came from Britain. A. H. Gibson and S. D. Heron of the RAE (Royal Aircraft Establishment) at Farnborough had learned, in the course of extensive

FIG. 6-2 Engine installation in Salmson S.A.L. 2s at the Salmson plant, 1917.
The small tear-drop-shaped device with the propeller, on the center landing gear
leg, is a wind-driven generator. These were common in all nations. They powered
the radio and landing lights until engines began to be equipped with generators
as well as magnetos. The fatigue shown in the workers' faces, after 3 years of
long wartime hours, is striking. *(Courtesy of Florence Lubin.)*

experiments aimed at improving the RAF (Royal Aircraft Factory) engines,[1] just what the design rules were for good air-cooled cylinders. They called for an aluminum head threaded onto a steel cylinder, widely angled valves set crosswise to the airflow, and plenty of cooling air passing between the intake and the exhaust ports (Fig. 6-3). These principles were enunciated in 1917 and were deviated from at peril right to the end of the radial era. Unfortunately, the first large British radials did deviate from them.

These first radials were the ABC Wasp and Dragonfly. Probably the basic trouble with the ABC engines stemmed from their designer's awareness that valves can't be allowed to seat directly in an aluminum head. Aluminum is just too soft. Either the head has to be made of iron or steel or else hard-metal inserts have to be somehow set in the aluminum head. Using inserts in an aluminum head is by far the better way, since the heat conductivity

of aluminum is much greater than that of ferrous metal; but, since aluminum also expands more with heat than does iron or steel, a way must be found to prevent the inserts from loosening up when the head gets hot. In the end, this problem turned out to be an imaginary one. A cast-in-place insert stays put very nicely, because the valve hammers it in firmly. However, as anyone who has ever coped with an S.U. carburetor knows, British engineering sometimes tends toward elaborate solutions for nonproblems.

The cylinder designed for the ABC radials by Granville Bradshaw was just such a piece of misplaced ingenuity. It was machined from solid, as were most rotary-engine cylinders, and the cooling fin design for the head was hardly more effective than those used on rotaries. Bradshaw's idea for keeping the cylinders cool was to plate them with copper. Copper is a better heat conductor than steel, but, as the heat had to pass through the steel before it got to the copper, it might just as well have been left off.

The Wasp came first and was the smaller of

---

[1]These were air-cooled V-8s and V-12s (see Chap. 9).

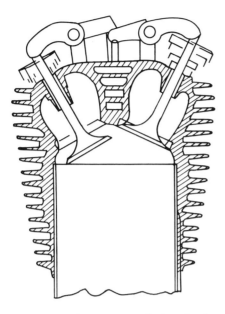

FIG. 6-3 1918 Gibson-type cylinder, the first air-cooled cylinder design capable of dissipating the heat of a large, high-output engine. The "volute" springs shown were popular for several years because they did not require as much height as coil springs.

the two. It had 7 cylinders of 4 ½-in. (114.3-mm) bore and developed 170 hp fairly frequently. The larger Dragonfly had nine 5 ½-in. (140-mm) cylinders and gave 360 hp on rare occasions. When the Wasp was tested, its temperamental behavior was not recognized as being symptomatic of faulty basic design. It is usual for promising new designs to need considerable debugging, and the feeling was that the Wasp would respond to a good program of troubleshooting. When it worked, the Wasp had the best power-to-weight ratio of any engine to date. If it could be straightened out, its design represented a great step forward.

Thus, when the government was offered the paper design of the proposed Dragonfly, it took the hook. Expecting great things from these untried engines, the government gave all other development work a lower priority and, in order to have manufacturing capacity available as soon as the designs were frozen, began phasing out contracts for the BR2 rotary. When the Dragonfly proved a

total failure, it looked for a while as though the British would run out of engines before they ran out of Germans, and there was something of a panic to reinstate Bentley production.

The Wasp was designed to produce 28.3 hp per cylinder, the Dragonfly 40. Even had the cooling arrangements for the Wasp been adequate, there should have been little cause for confidence that its design could be scaled up successfully. When a cylinder is scaled up, the power and heat rise as the cube of the enlargement factor but the heat dissipating ability only as its square. The Dragonfly ran so hot that it simply cooked its valves. The late S. D. Heron said that its consumption should be measured in pounds of valves per hour, rather than in pounds of fuel (Fig. 6-4).

A somewhat better radial, Cosmos Engineering's Jupiter, was developed at the same time that the Wasp and the Dragonfly were. It has been said that the Jupiter evolved from the ABC engines, but this is not true. Both were designed independently in 1917 and built in 1918. Had the government

FIG. 6-4 ABC Wasp, 170 hp, 1917. The four annular fins above the spark plugs were all the heads had to cool them. The mysterious-looking cable is a sling, retouched out above the engine proper. (Courtesy of NASM-Smithsonian.)

not backed the wrong horse, the Jupiter might have been ready in time to have seen combat.

The Jupiter was a 9-cylinder engine, larger, at 1753 cu in. (28.73 L), than the 1389-cu-in. (22.76-L) Dragonfly. Its cylinder head design was not a first-class one but was far better than that of the Wasp or Dragonfly. It was called a "poultice" head and consisted of an aluminum cap with fins, bolted to the flat top of the closed-end steel cylinder. The valves thus seated on steel, and the hot gases passed out through finned aluminum ports. For this combination to effectively conduct heat away from the valve seats, and for it to not leak at the ports, the fit between the cylinder and the cap had to be extremely good. Unfortunately, aluminum parts don't come back to quite their original size and shape after a cycle of heating and cooling, as many Chevrolet Vega owners have learned in recent years. Thus this critical fit deteriorated with use, and Jupiters suffered a noticeable drop-off in power as overhaul time neared. The Jupiter was

FIG. 6-5   Bristol Jupiter Radial, early to mid-twenties, with "poultice" heads. As with the Armstrong-Siddeley engines, the workmanship and finish were superb. *(Courtesy of Rolls-Royce, Limited.)*

widely used after 1918, but it was not really a first-class engine until a better head design was introduced with the F series in 1929 (Fig. 6-5).

The Jupiter had a 14-cylinder companion called the Mercury. Basically, a four-stroke radial must have an odd number of cylinders to make an even firing order possible. A 9-cylinder radial's firing order is 1, 3, 5, 7, 9, 2, 4, 6, 8. This occurs over two revolutions. Skipping every other cylinder each time, no. 1 is ready at the start of the next revolution. Accordingly, a radial with an even number of cylinders is always a two-row engine with a two-throw crankshaft. A six is really two threes built as one engine, a ten is two fives, a fourteen is really two sevens, and an eighteen is two nines. The two-row radial is both more complicated and heavier than a single-row engine, but it can make sense for two reasons. If more power is needed, a larger engine is normally required, and this means either larger cylinders or more cylinders of the same size. Larger cylinders mean a larger-diameter motor, possible structural and heat problems, and more vibration. Using more cylinders of a proven design tends toward "low-risk" technology—one pretty much knows that they will work—as well as resulting in a smoother engine and avoiding drag-producing increases in diameter.

The Jupiter and the Mercury were too late for war contracts, and Cosmos Engineering sold out to Bristol Aircraft.[2] With Bristol's resources behind it, the Jupiter never looked back. By the time the F-model head was introduced, its innards had been so highly perfected under the demanding leadership of Roy (later Sir Roy) Fedden that the later Jupiter was as fine a radial as any in the world. It was also one of the most extensively licensed for foreign manufacture, almost every country in the world having Jupiter licenses before the design came to the end of its evolution. First to take a license to build the Bristol engines was Gnome-Rhone in France, which badly needed a modern line to replace its rotaries.

In 1919 an even better radial, the Armstrong-Siddeley, was built. If the ABC engines represent

[2]Originally British and Colonial Aircraft, an offshoot of (believe it or not) the city of Bristol's trolley (tram) lines firm.

a classic case of incautious enthusiasm, the Armstrong-Siddeleys are at the opposite extreme. The design had actually originated at the RAE (Royal Aircraft Establishment) as far back as 1916. It was to have been the RAF 8, incorporating the findings of Gibson and Heron on proper cylinder head design for air-cooled engines. The RAE had been the RAF (Royal Aircraft Factory), but its manufacturing activities were phased out as a result of a Parliamentary investigation in 1917, and the proposed RAF 8 design ended up with Siddeley-Deasey, which became Armstrong-Siddeley. The fate of the RAF/RAE parallels that of the American Naval Aircraft Factory in Philadelphia. In both cases the aircraft industry was happy to build designs originated by the government facility but set up an outcry when the costs of government-built products proved to be lower than those from private industry. Congress and Parliament responded similarly; under the sacred banner of free enterprise, the government factories were forced to stop manufacturing and to confine themselves to research and development. A further refinement in the British case was that nothing happened as long as the RAF 8 design belonged to the relatively small Siddeley-Deasy firm, but government contracts came along when the large and influential Armstrong-Whitworth combine took them over.

The Armstrong-Siddeley radial, called the Jaguar, was a good design from the start. It used a short stroke design—the bore and stroke were nearly equal at 5×5.5 in. (127×140 mm)—to keep the diameter down, and it had a proper threaded-on aluminum head. The valves were splayed out at enough of an angle to allow plenty of cooling fins and adequate air passage between them. The aluminum crankcase was forged—not cast—and there was a blower at the back to slightly pressurize the manifold in the interest of good fuel distribution.[3] When Pratt and Whitney used these features on the first Wasp in 1926, its designers seem to have been surprised to learn that Armstrong-Siddeley had been ahead with both. Clearly, insularity is by no means a British monopoly!

[3]The Jupiter had a mixing chamber with spiral passages at the back, which helped but didn't do as much as a blower would have.

Armstrong-Siddeley actually built two engines around the same good cylinder design, the 7-cylinder Lynx of originally 150 hp and the 14-cylinder Jaguar of 300 to 320 hp. Their power-to-weight ratio was not as good as those of the dreaded ABCs, but, except for a spot of bother getting the crankshaft counterweights right, they were entirely reliable from the start. These engines were good enough to lead to a 10-year period during which the air-cooled radial was supreme for British fighter planes (Fig. 6-6).

The Rise of the Radial Engine in America

While England was trying to develop a radial for war use, government participation in engine development here was concentrated on the allegedly supernal Liberty. The emergence of the American radial was thus somewhat slower. As early as 1913

FIG. 6-6 Armstrong-Siddeley Jaguar, 14 cylinders, 360 hp, early 1920s. Note the very light, "skeletonized" rocker arms. The small diameter of this engine, when used in aircraft with bulky fuselages, led to some highly unaesthetic shapes. (*Courtesy of NASM-Smithsonian.*)

there had been an American radial called the Smith Static Radial that had attracted the attention of the British government. The Smith had 10 cylinders in the same plane but used a two-throw crankshaft by means of an odd expedient. The connecting rods of cylinders 1, 3, 5, 7, and 9 slanted forward to engage the first throw, while those of 2, 4, 6, 8, and 10 sloped back to drive the second throw. This atrocious piece of engineering passed bench tests and was actually flown, but its deficiencies must have shown up early in the game, because it had been dropped by 1917.

The practical radial in the United States was at first largely the work of Charles L. Lawrance. Lawrance's small company had begun unpromisingly with a terrible opposed twin, usable only in "Penguin" nonflying trainers (see Chap. 8), and had then gone on to a rather decent 3-cylinder radial of 60 hp. Using the same cylinders, Lawrance then built a 9-cylinder radial of 180 hp in 1921. This engine, the J-1, evolved eventually into the immortal Wright Whirlwind. The lower end was good enough to remain largely unchanged during the whole development of the engine, the pistons were daringly short for the time, and the cylinders were at least partly correct (Fig. 6-7). These were aluminum with steel sleeves, and while the head design needed development, it was adequate to cope with an output that early went to 200 hp. The Navy, already interested in air cooling, bought 200 J-1s for use as training planes.

Lawrance, though, lacked the necessary capital to exploit his promising design. Meanwhile, Wright had been building Hispano-Suiza engines under license and had gotten very good at it. It had even improved on Birgikt's design by using open-ended sleeves and inserts in the aluminum head, but its efforts in the radial field were not working out well. During this critical period the Navy's powerplant chief, Lt. Com. Eugene V. Wilson, was in the key role, and he played it well. By exerting pressure on Frederic B. Rentschler, who had created Wright Aeronautical from the former Wright-Martin, and on Lawrance, he instigated a merger between Wright and Lawrance.

Wright had capital and facilities and, in Messrs. Wilgoos and Mead, excellent engineers; but it was a third addition to the mix that made the

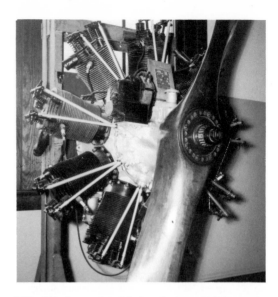

FIG. 6-7 Lawrance J-2, 9 cylinders, 200 hp, 1923. Head and upper part of cylinders were aluminum. Both intake and exhaust were at the rear, allowing unobstructed flow of cooling air to heads and cylinders. Compare the spark plug placement with that on later engines. *(Courtesy of the author, taken at Glenn Curtiss Museum, Hammondsport, New York.)*

Whirlwind the great engine that it soon became. Samuel D. Heron, Professor Gibson's assistant at Farnborough, had left England and was working for the U.S. Army Air Corps at Wright Field. When Heron joined Wright, the firm got the services of a man who knew more than anyone else in the world about the design of air-cooled cylinders. With the head that Heron designed for the new J-5 model, horsepower went to 220 and the highest reliability of any aircraft engine to date was achieved. The Whirlwind had arrived. This head, shown in Fig. 1-1(b), would still look modern 20 years later. The valves were even more widely angled than the Jaguar's, allowing them to be larger and permitting very generous finning and ample airflow between the ports. For the first time, a radial had all the air it needed and had it in exactly the right place.

Wright paid Giuseppe Bellanca to design the Columbia as a demonstration craft for the J-5, and it soon brought the engine to the attention of the aviation world by setting a new endurance record. This focused Charles Lindbergh's attention on the Whirlwind-Bellanca combination, and he tried hard to buy it for his transatlantic attempt. However, Bellanca and the Columbia were by then owned by Charles Levine, who was willing to take Lindbergh's money but not to let him have the airplane, and so Lindbergh went to Ryan for his plane. The Spirit of St. Louis was designed around the J-5, but in one respect, Lindbergh's Whirlwind was nonstandard: the rocker arm lubrication had been modified to allow for longer-than-normal periods of continuous operation (Fig. 6-8).

If, in 1927, there remained any doubts about the future of the air-cooled radial, Lindbergh's flight dispelled them. It was well known that Hawker and Grieve and Alcock and Brown had experienced problems with their Rolls-Royce engines on their transatlantic ventures. By contrast, Lindbergh's and, within the next few weeks, Chamberlain's and

Byrd's air-cooled Whirlwinds never gave them a moment's concern.[4]

It is not easy today to appreciate the impact that the Whirlwind had on the aviation world in 1927. Transatlantic flight difficulties with regard to engine reliability and efficiency were so great that only Nungesser and Coli's and Fonck's fiascos had come after the events of 1919. Then suddenly the Whirlwind made it look easy. Yet commitment to the air-cooled radial was less than total at, of all places, Wright. Wright was moving in all directions at once in 1926. It had a homegrown radial with poultice heads, the R-1: a monster water-cooled six for dirigibles; the T-2,[5] a 500-hp replacement for

---

[4]Though Kingsford-Smith had (significantly) rocker arm troubles a little later on his flight to Australia.

[5]T-2s were tried on torpedo planes. They were better than the Packards, but their reliability was below that of the radials.

FIG. 6-8   Charles Lindbergh, Ryan NYP type, license NX211 and the Wright J-5 Whirlwind—the most famous team in aviation history. Note that the engine installation has had more attention paid to streamlining than most engines received before the coming of outside cowlings for radials. The white objects visible on the rocker arm covers are special grease fittings with reservoirs and tiny spring-loaded pistons, whose job was to gradually force lubrication into the rocker arm bearings during the long flight. These replaced the standard Alemite fittings. (Courtesy of Everett Cassagneres, Ryan Aircraft Historian.)

the Liberty; a cast block to put on the Liberty base instead of the welded cylinders; a marine engine; and the license-built Hispano-Suizas. The directors seemed to feel that the future lay with water cooling, insofar as they had any sense of direction at all. As a result of this internal discord, Wright's president, Frederic B. Rentschler, left the firm in 1926 and started a new one dedicated strictly to the large radial.

Again Wilson and the Navy had a great deal to do with this move. One-third of all the Navy's engine problems at the time were traceable to water-cooling system difficulties. The Navy wanted an air-cooled engine in the 400-hp class to replace D-12s in fighters and, if possible, one of over 500 hp to use instead of Packards in torpedo planes and patrol boats. Wright's first try at a large engine, the Simoon, was a failure—the triumphs of the Cyclone were still in the future.

Rentschler, then, put together an assurance from Navy sources that they would buy all the good, big radials he could make, much of the design staff of Wright, led by George J. Mead, and financing from a well-known Hartford, Connecticut, machine-tool builder, Pratt and Whitney—which also gave its name to the venture. The resulting engine, developed almost as rapidly as the Liberty, was the 1340-cu-in. (22-L) Wasp, and it was a marvelous piece of work. More than a mere enlargement of the Whirlwind, it had a forged aluminum crankcase, enclosed and continuously lubricated rocker arms, a vibration damper on the crankshaft, and, more for good fuel distribution than for improved altitude performance, a low-pressure blower. As mentioned earlier, some of these features had been previously used by Armstrong-Siddeley, but the Pratt and Whitney team had arrived at them independently.

It is interesting to compare the world's three best radials of 1927. At 2000 rpm, the Jaguar was rated at the highest speed of the three (entirely appropriately, since it had been around longest and had been the beneficiary of more development work). The others were slightly slower at 1850. The Jaguar produced 1 hp/4.2 cu in. (14.5 hp/L) of displacement, the Whirlwind 1 hp/3.57 cu in. (17.1 hp/L), and the Wasp 1 hp/3.16 cu in. (19.3 hp/L). Their weight-to-power ratios were 2.53

lb/hp (1.15 kg/hp) for the Jaguar, 2.16 lb/hp (0.98 kg/hp) for the Whirlwind, and 1.53 lb/hp (0.695 kg/hp) for the Wasp. The significance of the specific-output figures lies in the fact that the Whirlwind worked at higher cylinder pressure (BMEP) than did the Jaguar and the Wasp worked at still higher cylinder pressures. As anyone who has ever been involved with hopping up an engine knows, the problem is not so much getting more power out of an engine as it is getting the engine to stand up to the higher output. It can therefore be assumed that these figures represented the highest power that the respective engines could handle in terms of heat dissipation and of bearing durability, all of which indicates that the Whirlwind's head design was a distinct improvement over the Jaguar's, and the Wasp's was a still further improvement on the Whirlwind's.

The Navy had first call on Wasp production (Fig. 6-9) but allowed some civilian buyers to step ahead of line and get some of the early ones. When the Wasp was substituted for the center Whirlwind

FIG. 6-9   The first Pratt and Whitney Wasp ever built, 1926—the engine that completed the radial revolution begun by the Whirlwind. This first engine was actually too precisely built; the first endurance test run was stopped and the engine torn down to make the piston rings a looser fit because the oil consumption was too low! (Courtesy of Pratt and Whitney Division, UTC Corporation.)

in the Ford Trimotor, its extra power, in the words of one of its designers,[6] "made an airplane out of it." It produced sensational results in the Lockheed Vega. The Vega needed the Wasp's added power, even more than the Ford did, to realize the performance potential of its clean, strong airframe. Wasp-powered, the Vega became the dominant second-generation (considering the Bellanca as the first) single-engine speed and distance aircraft.

As soon as the Wasp's design was complete, Pratt and Whitney's George Wilgoos began work on a larger engine. This was the Hornet, of 575 hp; at 1860 cu in. (30.5 L) it was 38 percent larger than the Wasp. With a Hornet installed in place of the Packard in the Martin T4M series, a 1-ton (909 kg) torpedo could be carried for a range of 665 mi (1070 km). The torpedo plane was, at last, a practical proposition. In a bulky-fuselage large single-engine plane—and the Martin was large indeed (I remember the ones the Reserve had at Floyd Bennett Field in the early thirties as dwarfing everything in sight)—the considerable diameter of a maximum-size 9-cylinder radial was not objectionable. The 56 ¾-in. (1440-mm)-diameter Hornet looked right in the Martin. In contrast, the British Vickers Vildebeeste, a torpedo plane comparable in design, dimensions, and performance to the T4M, used the 14-cylinder Jaguar Major of only 46.45-in. (1180-mm) diameter and looked decidedly odd with it.

The Hornet, while a vast improvement on the Packard, was not quite the trouble-free engine that the Wasp was. Careless throttle handling could cause piston dishing and other problems. A better big nine of the same era was the Wright Cyclone. Wright had reacted vigorously to the departure of the group that went to make up the nucleus of Pratt and Whitney. Now with Lawrance in the president's chair, it dropped the deadwood, redesigned the Whirlwind to 975 cu in. (15.98 L) and 300 hp as the J-6-9, and introduced the Cyclone. At 1750 cu in. (28.7 L), the first Cyclone was a little smaller than the Hornet and developed 25 hp

less. The relative success of the two large rivals can be judged from the subsequent development of both: the Cyclone was upsized to 1820 cu in. (29.8 L) and enjoyed many years of success, while the Hornet, reduced to 1690 cu in. (27.7 L), became a very fine engine but fell into a less popular size and power class. Pratt and Whitney's great success in the 1800-cu-in. (30-L) class came with a 14-cylinder engine, the later R-1830.

Military acceptance of the Cyclone was at least as good as it was for the Hornet, and civilian use was probably greater. By 1929, not only had Curtiss merged with Wright: in addition, Rentschler had put together the United Aircraft combination. Now Curtiss had an in-house source for modern engines and Wright had an in-house customer. At the same time, Pratt and Whitney could count on having the inside track for the engine requirements of Boeing, Stearman, Chance Vought, and Sikorsky. Any attempt to gauge the preference between the Hornet and the Cyclone among aircraft manufacturers, therefore, should be based on use by manufacturers outside the two great combines. In 1930, by which time the qualities of both engines had become reasonably well known, the two large radials were being used by an equal number of manufacturers. By 1933—significant as the year in which Pratt and Whitney announced the 14-cylinder R-1830 Twin Wasp as its new large radial—the Cyclone had established a noticeable dominance. In particular, the two most significant new planes of the year, the Douglas DC-1 and Martin's fast B-10 bomber, were both designed for the Cyclone. It would seem that Pratt and Whitney developed the R-1830 as its new large radial and, in a sense, Hornet replacement, not only because of its own professional dissatisfaction but because the Hornet was not finding a satisfactory market.

Wright had made another shrewd move in 1927. As popular as the original J-5 Whirlwind was, it was already obsolete in a sense. There was no longer any need for a 220-hp engine to be burdened with the cost and complexity inherent in 9 cylinders. Accordingly, Wright designed a completely new engine, still called the Whirlwind but very different from the original, of 975 cu in. (15.98 L) and 300 hp. Known as the J-6-9, it was up to as much as 450 hp with satisfactory reliability by

[6]John G. Lee. Contrary to popular opinion, William B. Stout did not design the Trimotor; in fact, Ford had to fire him to get the job done.

1930 and forced Pratt and Whitney to respond with a new design of its own, of which more later (Fig. 6-10).

American popular engine nomenclature was undergoing a change around 1930. Up to that time, we had been content to refer to engines by their makers' given names—Wasp, Whirlwind, Challenger, or whatever. However, the American love for numbers, so odd to foreigners—or was it a desire to seem to be on the inside of things?—led to a preference for numerical designations. Thus, the J-6 was always a J-6 and never a Whirlwind. After about 1932, all hands began to use the military designations, based on displacement, in ordinary conversation and writing. It was not universal; the J-6 was seldom called an R-975, and the Wasp Junior was usually called by its name. But the bigger engines were generally known by their military designations.

FIG. 6-10 Wright J-6-9 Whirlwind, 300 hp, 1928. Completely redesigned from the original Whirlwind J-5—larger, cam ring and pushrods at the rear, magnetos in back, automatic rocker arm lubrication, etc. Various models had different exhaust collector rine locations; the frontal position is not an infallible recognition feature. *(Courtesy of NASM-Smithsonian.)*

France and Germany Awaken to the Radial

There was little original work done with large radials on the Continent. France's one truly indigenous radial, the Salmson, was, it is true, available in versions giving as much as 500 hp, but it was little used. The most popular French radial was the Bristol-based Gnome-Rhone. It is just barely possible that, had Capt. Rene Fonck been a different sort of person, Gnome Jupiters might have powered the first New York – Paris flight. Three Gnomes were used in a large Sikorsky in which, as a result of ill-conceived gadgetry, Fonck crashed and burned while taking off on his attempted crossing[7] in 1926. The German Siemens was derived from the Jupiter, the BMW came from Pratt and Whitney designs, and the Hispano-Suiza radials were license-built Whirlwinds.

By 1930 or 1931, however, Gnome-Rhone had designed a line of radials of its own. Gnome used the same cylinder assembly in a five, the Titan of 270 hp; a seven, the 330-hp Titan Major; a nine, the Mistral of 550 hp; and the 18-cylinder Mistral Major of 800 hp. They were all quite American in appearance except for a certain slenderness of the cylinders, caused by the bore being a little smaller in relation to the stroke than was common American practice. As seen from the front, each pair of pushrod covers converged as they approached the crankcase to enter the case directly behind each other. This was caused by Gnome's use of two separate cam rings instead of one cam ring with separate intake and exhaust lobes, as was common American practice.

These were good engines, widely licensed abroad and much used by Air France and the French military. Nevertheless, both military and commercial aviation in Europe were more oriented toward liquid cooling than was the case in Britain and the United States. Searching out the reason becomes a case of the chicken and the egg. Was there relatively little development of radials because the liquid-cooled engine had the inside track, or were liquid-cooled engines preferred because of a dearth of development work on radials? It is prob-

[7]Fonck and one companion survived; two others did not.

able that the choice would favor liquid cooling when academically trained people made the decision, as seems to have been the case on the Continent. Liquid-cooled engines used less fuel and made for better streamlining; on paper they would have the advantage for many uses. Acceptance of the radial, on the other hand, suggests a more pragmatic approach based on experience with reliability and ease of servicing.

British Designs of the Late Twenties

The Jupiter F came out in 1929 with a head that at last cooled well, making it England's best radial engine. It was available with outputs as high as 575 hp by 1931. The Jupiter's new head was a great advance over the old poultice head, but it was unduly complicated. It employed four valves per cylinder, exposed to the air and operated by a complex arrangement of rocker arms, and had a device for maintaining constant tappet clearance regardless of cylinder heat and expansion. Since there is no evidence that all this worked any better than the far simpler arrangements on most other radials, it was a splendid example of finding answers to questions that were not being asked. A look at Fig. 6-11, showing the Jupiter's smaller companion, the Titan, will help the assiduous reader understand it.

The rear pushrod of each cylinder rocks a pair of levers on short, hollow shafts, on the back of which are other levers that actuate the two exhaust valves. These hollow shafts can be described as Z-type rocker arms. Through each one runs another, longer shaft, also equipped with levers. The outer pushrod operates the front pair of these levers, and the rear pair actuates the intake valves at the rear of the cylinder heads. Each of the four valves of each cylinder thus has its own "Z-type" rocker arm. The other two rods at the front of each cylinder constitute the temperature compensation. the front bearing of each pair of coaxial Z-type rocker arms is attached not to the cylinder head but to one of these steel rods. As these rods are in the cool incoming air, as are the nearby pushrods, they stay the same temperature

FIG. 6-11   Bristol Titan, Series F, about 250 hp, 1931. The improvement in head cooling over the earlier poultice head is obvious and the line of demarcation between the threaded-on aluminum head and the steel cylinder can be clearly seen; the aluminum did not come down as far over the steel as it did on American engines. The small hex nuts with a spot in the middle, visible at the front of each rocker arm shaft, are Tecalemite grease fittings; the Tecalemite system demanded more care in use than even the Zerk and was far less positive than the Alemite. Also very clear is the extremely good finish and workmanship on this overly complex engine. (*Courtesy of NASM-Smithsonian.*)

and length as the pushrods, and the tappet clearance doesn't change. It's simple enough when it's explained, isn't it?

Incidentally, notice the spiral hooks on the shaft nose. Those readers who have worked with pre-war automobiles will recognize this as the engagement device for a starting crank and wonder what an aircraft engine is doing with such an attachment. During the First World War and into the twenties, some use was made of a device called a "Hucks Starter," involving a shaft mounted on the back of a light truck and driven off a power takeoff. The truck would be backed up to the aircraft, the shaft engaged with the propeller, and the power takeoff engaged to turn the propeller. Since this involved some danger of damage to the propeller, despite the use of padded jaws on the end of the shaft, a crank engagement device on the nose of the propeller shaft, called a "Hucks Starter Dog," came into use.

The Armstrong-Siddeleys were refined as time passed but did not change in their basic design from what they had been in 1919. The line was an extensive one by 1930. It included 5-cylinder Genets of 88 and 110 hp; the 5-cylinder Mongoose (a mongoose is a cat?) of 165 hp; and the 7-cylinder Lynx and Cheetah (ex-Lynx Major) of 235 and 280 hp, respectively, for low- and medium-power applications. For higher-power needs there was the 14-cylinder Leopard of 815 hp, the Jaguar (a double Lynx) of 510 hp, and the Panther (ex-Jaguar Major) of 525.

Around 1929 the British and, to some extent, Curtiss-Wright in America began to use radials in export versions of aircraft that had originally been designed for liquid-cooled engines. Invariably this change, probably made to lower cost and improve servicing, led to lowered performance if the same power was used. The Hawker Horsley torpedo plane required an 805-hp Leopard to obtain the same speed that was conferred by a 760-hp Rolls-Royce Condor in the domestic version. The coming of the Townend Ring and the NACA cowl reduced this differential but did not eliminate it. Nevertheless the late twenties saw an inexorable trend toward air-cooled radial power wherever reliability, safety, and overall economics took precedence over the need for maximum performance.

This was partly due to a new factor that became more important around 1930. Short-term—usually 5-min—power was becoming an important parameter. As early as 1920, a British writer had pointed out that the power characteristics needed in civilian service were quite different from those of military craft. Military craft are generously powered because of combat requirements, and this ample power makes takeoff no problem. Transports tend to be relatively more heavily loaded and less powerful. They need full power only for takeoff; once in the air they can afford to throttle back

more than military planes can. This set of facts began to show up in the published power ratings of European engines in the mid-twenties, but it was not until about 1933 that every major American engine carried a "maximum for takeoff" rating at higher-than-normal power.

We did not need such a rating until controllable-pitch propellers, retractable wheels, and wing flaps became common. Despite what was said earlier, early transport planes like the Fokker and Ford trimotors did not need extra engine power to get into the air. Because they were so poorly streamlined, they had to be fairly generously powered in relation to their wing loadings if they were to have an acceptable cruising speed for American distances. A Ford, for instance, can land or take off in less than 750 ft (229 m) at sea level.[8] In contrast to this, a plane like a DC-3 could slip along nicely at a rather moderate cruise power setting, and so it didn't need big power once it had climbed out, but it did need to be able to call on the engines for plenty of power for takeoff. It has been said that the landing makes the heaviest demands on the pilot but the takeoff makes the heaviest demands on the aircraft.

The point of all this, historically, is that it becomes increasingly difficult to describe engines consistently in terms of their power rating. Over and over again one finds seemingly contradictory statements about the power output of certain engines. The fact is that, after 1928, more and more engines did have several power ratings at the same time. One can say of the J-5 that it was a 220-hp engine, but no such straightforward statement can be used for later engines.

Another change in the radial engine picture began to become important after about 1928. The range of available power began to become wider, and the large engines began to diverge in design characteristics from those used for lightplanes. To do justice to this divergence, the next chapters will consider the major radials separately from the smaller ones.

---

[8] If you doubt this, go to Port Clinton, Ohio, and buy a ticket to Put-in Bay on Island Airways' 5-A-T.

Data for Engines

Maker's name	Date	Cyl-inders	Config-uration	Horse-power	Revolutions per minute	Bore and stroke, in. (mm)	Displace-ment, cu in. (L)	Weight, lb (kg)	Remarks
Salmson M-7	13	7	R, L	90	1250	4.7 (120) × 5.5 (140)	765 (12.5)	375 (170)	
Salmson M-9	14	9	R, L	140	1300	4.7 (120) × 5.5 (140)	868 (14.2)	465 (211)	Weight includes radiator
Salmson Z-9	16	9	R, L	250	1550	4.9 (125) × 6.7 (170)	1146 (18.8)	473 (215)	Late Z-9s had master-and-link rods instead of Canton-Unne system
Fiat A-18	17	9	R, L	320	2000	5.1 (130) × 5.9 (150)	1094 (17.9)	NA	
ABC Wasp	16	7	R, A	170	1800	4.5 (114) × 5.9 (150)	667 (10.9)	290 (131)	Weight?
ABC Dragonfly	17	9	R, A	320	1650	5.5 (140) × 6.5 (165)	1389 (22.8)	656 (298)	Design work begun 1913
Cosmos Jupiter	18	9	R, A	450	1800	5.8 (146) × 7.5 (190)	1752 (28.7)	700 (318)	First Bristols same
Armstrong-Siddeley Jaguar	19	14	R, A	400	1700	5 (127) × 5.5 (140)	1512 (24.8)	850 (386)	Mark IVc gave 490 hp, had enclosed valve gear
Armstrong-Siddeley Lynx	20	7	R, A	215	1900	5 (127) × 5.5 (140)	756 (12.4)	512 (233)	Still 215 hp in 1936
Smith Static Radial	14	10	R, A	150	1250	4.5 (114) × 5.5 (140)	875 (14.3)	372 (169)	Made only in experimental quantities
Lawrance L-3 and L-5	19	3	R, A	65	2000	4.3 (108) × 5.3 (133)	223 (3.7)	175 (97.5)	Similar L-4 called Wright Gale
Lawrance J-1	22	9	R, A	200	1800	4.5 (114) × 5.5 (140)	787 (12.9)	476 (216)	
Lawrance J-2	23?	9	R, A	NA	NA	4.9 (124) × 5.5 (140)	924 (15.1)	NA	Only two built
Wright J-5	25	9	R, A	220	1800	4.5 (114) × 5.5 (140)	788 (12.9)	510 (232)	
Wright R-1	20	9	R, A	350	1800	5.6 (142) × 6.5 (165)	1454 (23.8)	NA	
Wright Simoon (R-1200)	25	9	R, A	350	1900	5.5 (140) × 5.5 (140)	1176 (19.3)	640 (291)	First to use R- followed by displacement designation system

Data for Engines (continued)

Maker's name	Date	Cyl-inders	Config-uration	Horse-power	Revolutions per minute	Bore and stroke, in. (mm)	Displace-ment, cu in. (L)	Weight, lb (kg)	Remarks
Pratt and Whitney Wasp	26	9	R, A	425	1900	5.8 (146) × 5.8 (146)	1344 (22.2)	650 (295)	
Pratt and Whitney Hornet A	26	9	R, A	525	1900	6.1 (156) × 6.4 (162)	1690 (27.7)	800 (362)	C, D and E similar; maximum 875 hp
Pratt and Whitney Hornet B	29	9	R, A	575	1950	6.3 (159) × 6.8 (171)	1860 (30.5)	860 (390)	Piston troubles led to return to A size
Wright Cyclone A thru Cyclone E	27	9	R, A	525	1900	6 (152) × 6.9 (175)	1750 (28.7)	760 (345)	
Wright Cyclone F	30	9	R, A	575	1900	6.1 (156) × 6.9 (175)	1820 (29.8)	974 (443)	Weight may be for G type
Wright J-6-9 Whirlwind	29	9	R, A	300	2000	5 (127) × 5.5 (140)	975 (16)	550 (250)	This is 1929 hp: later models gave up to 450 hp
Siemens Early		See Bristol Jupiter							
Gnome-Rhone Titan 5Kds	28	5	R, A	260	2000	5.8 (146) × 6.5 (165)	843 (13.8)	473 (215)	All these Gnomes had same 146 × 165 mm cylinders
Gnome-Rhone Titan Major 7Ksd	28	7	R, A	370	2000	5.8 (146) × 6.5 (165)	1180 (19.3)	787 (358)	Original output was 550 hp
Gnome-Rhone Mistral 9Ktr	28	9	R, A	770	2410	5.8 (146) × 6.5 (165)	1517 (24.9)	1111 (505)	
Gnome-Rhone Mistral Major 14 Kbs	28	14	R, A	705	2000	5.8 (146) × 6.5 (165)	2360 (38.7)	NA	Maximum power was 800 at unknown rpm
Bristol Jupiter VI FS	29	9	R, A	465	1870	5.8 (146) × 7.5 (190)	1752 (28.7)	775 (352)	VI FM gave 500 hp
Bristol Titan III F	29	5	R, A	240	1870	5.8 (146) × 6.5 (165)	824 (13.5)	775 (352)	
Armstrong-Siddeley Genet	26	5	R, A	82	2200	4 (102) × 4 (102)	251 (4.1)	203 (92)	
Armstrong-Siddeley Mongoose	26	5	R, A	150	1850	5 (127) × 5.5 (140)	540 (8.9)	340 (155)	165 hp for late models
Armstrong-Siddeley Cheetah	30	7	R, A	295	200	5.3 (133) × 5 (127)	834 (13.7)	556 (253)	
Armstrong-Siddeley Leopard	28	14	R, A	800	1700	6 (152) × 7.5 (191)	2969 (48.9)	1637 (744)	Most powerful radial in world when introduced
Armstrong-Siddeley Panther	29	14	R, A	525	NA	5.3 (133) × 5.5 (140)	1829 (30.0)	980 (451)	Panther X, 1936, gave 752 hp

R—radial; A—air-cooled; L—liquid-cooled; NA—not available. Open figures—U.S. Customary; figures in parentheses—metric. Weights are dry.

SEVEN

The major radials.

Bristol and the Sleeve Valve

With one exception, the radial engines that fought the Second World War and that dominated air transport up to the coming of the jets all conformed to the design patterns that had evolved by the early thirties. That one exception was the sleeve-valve engine as developed in Britain by Bristol.

In a sleeve-valve engine, the pistons operate in sleeves which themselves move within the cylinders. In these sleeves there are holes that uncover the intake and exhaust ports at the proper times in the operating cycle. There are two significant types, the Knight and the Burt-McCollum. In a Knight engine there are two sleeves within the cylinder and the piston rides in the inner sleeve. An eccentric shaft replaces the conventional camshaft, and small connecting rods attached to this shaft move the sleeves up and down. In one alignment of the sleeves the exhaust port is uncovered, and in another, the intake is uncovered. Knight engines were used on the American Willys-Knight, the Belgian Minerva, and the English Daimler cars, but I know of no aircraft applications. Its only advantage lay in its mechanical quietness in a day when noisy tappets were common, and this had to be paid for with hard starting in cold weather, poor performance, and heavy oil consumption.

The Burt-McCollum is also known as the "single-sleeve" type. Its one sleeve turns with a wristlike motion in addition to moving up and down (Fig. 7-1). This allows the same hole in the sleeve to uncover both the intake port and the exhaust port in their turn. Few cars—the Scottish Argyll was one—used it, but it did promise better cooling and less friction than did the Knight type if its problems could be solved.

In 1927 Bristol, under Roy Fedden's leadership, undertook to solve those problems. If they could be solved, the resulting engine promised to be an improvement over the conventional type in three areas. First, pumping efficiency and fuel economy should be increased. This is because the conventional poppet valve is in the way of gas flow, like an adenoid, when it is open. In contrast, the opening in the sleeve offers a clear passage to the incoming mixture or escaping exhaust. Second, higher compression could be used, increasing power and again decreasing fuel consumption. Knock and detonation, which limit the compression ratio that can be

113

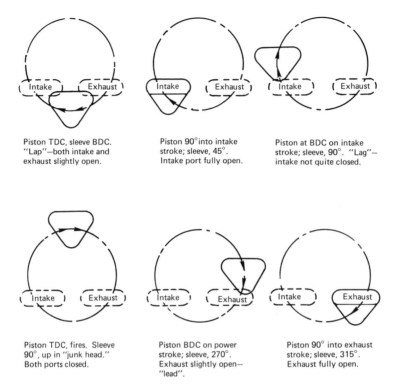

Piston TDC, sleeve BDC. "Lap"—both intake and exhaust slightly open.

Piston 90° into intake stroke; sleeve, 45°. Intake port fully open.

Piston at BDC on intake stroke; sleeve, 90°. "Lag"— intake not quite closed.

Piston TDC, fires. Sleeve 90°, up in "junk head." Both ports closed.

Piston BDC on power stroke; sleeve, 270°. Exhaust slightly open— "lead".

Piston 90° into exhaust stroke; sleeve, 315°. Exhaust fully open.

FIG. 7-1   Action of the opening in a single-sleeve-valve engine in opening and closing ports. This is drawn "developed"—as if the cylinder were unwrapped and laid flat. TDC—top dead center; BDC—bottom dead center; piston is at top and bottom, respectively, of stroke.

used with fuel of a given octane rating, are associated with the presence in the combustion chamber of the hot exhaust valve. Since there is no such valve in the combustion chamber of a sleeve-valve engine, compression ratios of up to one number higher can be used. Third, in the absence of the conventional overhead-valve gear, engine diameter can be reduced with beneficial results in lessened drag.

Bristol showed admirable persistence in pursuing its work on the sleeve-valve concept. It was not until 1933, 6 years after it had begun experimenting, that its first sleeve-valve radials passed their type tests and were put on the market. The first engines of the new type were the Perseus of 580 hp and the Aquila of 365. As befits a brand-

new design, these initial ratings were very conservative and rose to 810 for the Perseus and 500 for the Aquila by 1936. The figures for the original Perseus exactly match those of Bristol's poppet-valve Mercury of the same year, suggesting that Bristol may have given some thought to preserving the market for its more conventional engines. Also, Bristol could not have met the demand if the sleeve-valve line had become highly popular. British manufacturing technology was not up to producing the sleeves in quantity and on an interchangeable basis until almost the beginning of the war. At first, each sleeve had to be individually fitted to its own cylinder and piston. Likewise, there was, at first, no decrease in diameter for the sleeve-valve models as compared with the conventional engines. Exploitation of the advantages of the new type was for the future. In the beginning, it was enough that they prove themselves reliable, economical, smooth-running engines.

Development continued, and the Perseus reached 930 hp by 1939, but the real future of the sleeve valve lay with still larger engines. The Aquila found little employment, but its cylinders formed the basis for Bristol's next step up, the 14-cylinder Taurus. The Taurus, in various marks between 1010 and 1065 hp, powered some of the bombers that began the war with the RAF. It demonstrated the effectiveness of the sleeve-valve idea in the Beaufort bomber; when a Pratt and Whitney R-1830 of 1050 hp was substituted for the Beaufort's original 1010-hp Mark II Taurus, performance decreased slightly despite the Twin Wasp's lighter weight and higher output.

Perseus cylinders were then used for the 14-cylinder Hercules of 1375 hp. The major wartime mark, built in considerable numbers, was the Hercules VI of 1650 hp. It was used in the Lancaster II bomber in 1942 and in Bristol's own very successful twin-engine fighter, the Beaufighter (Fig. 7-2).

The final achievement in this great line of engines was the Centaurus, an 18-cylinder outgrowth of the Perseus. This was Britain's most powerful radial ever, giving 2520 hp in the Mark V and Mark VI, 2500 in the 1942 Mark VII, and 2470 in the post-war civil type 57. The Centaurus barely made it into the war, being apparently used only in a few Warwicks, fabric-covered scaled-up Wellingtons, but it came into its own in the immediate postwar era. It powered Britain's last piston-engined fighter, the Hawker Tempest Mark II, and some of the misconceived Brabazon transports on which so much effort was wasted.

Probably no more efficient radials than the big Bristols were ever built. The Perseus, for example, had a specific consumption of 0.43 lb (0.195 kg) at lean-mixture cruise, using 87-octane fuel. It is regrettable that there are no Burt-McCollum engines made for automobiles in these days of high-priced low-octane gasoline. A single-sleeve-valve engine ought to be able to deliver 5 to 10 percent better gas mileage than a comparable poppet-valve engine can, other things being equal.

When the sleeve-valve engines arrived, Bristol's poppet-valve engines seemed at first to go into a decline, but they soon recovered, stronger than ever. The Neptune and the Titan had already been dropped after 1930, and 1933 was the venerable Jupiter's last year. The last Jupiter had up to 595 hp available for takeoff—about 50 percent more than the first models—but this was both a lot for the engine and not enough for the market. Leaving the smaller-radial field to Armstrong-Siddeley, Bristol replaced the Jupiter with the Mercury and added the new Pegasus of 1753 cu in. (28.73 L); both were 9-cylinder jobs. In these engines the valve gear, while still working on the same linkage as the F-type Jupiter, was somewhat cleaned up. The temperature-compensating tie rods at the front were gone, the two pushrods were now enclosed in a single tube, and a dustcover enclosed the front parts of the rocker arms. The rear parts and the valve springs continued to be exposed.

The Pegasus eventually achieved 1000 hp. Its reliability, in 690- and 750-hp models, was one of the attractions of the Fairey Swordfish, that extraordinary survival of 1931 design practice that served the British as their only carrier-based torpedo plane throughout most of World War II and with which, in the absence of serious opposition, they accomplished so much. Mercuries, in pre-war aircraft that were still on hand in 1939, also saw some wartime service. Their most significant applications were in the Blenheim and the Blackburn Skua. The Blenheim was the first British *Schnellbomber,* suicidally obsolete by the outbreak of the war; the Skua was the first British low-wing, retractable-gear naval dive-bomber, corresponding roughly with our own Vought SB2U. However, by the time the war was well under way the uses for a 1519-cu-in. (24.9-L) engine were few, and those few could be better handled by the sleeve-valve Perseus.

While Bristol was going from strength to strength the Armstrong-Siddeley line, prominent up to the early thirties, was slowly dying. The last notable military application of Armstrong-Siddeley radials was in the Armstrong-Whitworth Whitley, Britain's first retractable-gear low-wing heavy bomber. Even here, while the Mark I used 920-hp Tiger VIIIs, the only version to see wartime combat

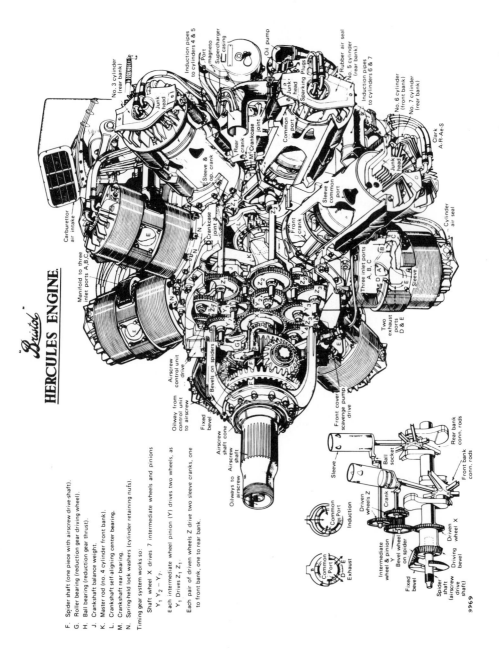

"Bristol" **HERCULES ENGINE.**

No. 3 cylinder (rear bank)

Induction pipes to cylinders 4 & 5

Port magneto

Supercharger casing

Oil pump

Junk head

Common port

Sparking Plugs

Rubber air seal

No. 5 cylinder (rear bank)

Induction pipes to cylinders 6 & 7

No. 6 cylinder (front bank)

No. 7 cylinder (rear bank)

Junk head

Clark A-R-Ae-S

Sleeve & op. crank

Rear crank

M'Crankcase joint

Sleeve common port

Front crank

Cylinder air seal

Carburettor air intake

Manifold to three inlet ports A,B,C.

Crankcase joint 3

Airscrew control unit drive

Oilway from control unit to airscrew

Fixed bevel

Airscrew shaft cone

Airscrew shaft

Oilways to Airscrew airscrew

Bevels on spiders

Three inlet ports A, B, C

Two exhaust ports D & E

Sleeve

F. Spider shaft (one piece with airscrew drive-shaft).
G. Roller bearing (reduction gear driving wheel).
H. Ball bearing (reduction gear thrust).
J. Crankshaft balance weight.
K. Master rod (no. 4 cylinder front bank).
L. Crankshaft self-aligning center bearing.
M. Crankshaft rear bearing.
N. Spring-held lock washers (cylinder retaining nuts).

Timing gear system works so:

Shaft wheel X drives 7 intermediate wheels and pinions

Y_1 Y_2 – Y_7.

Each intermediate wheel pinion (Y) drives two wheels, as

Y_1 Drives Z_1, Z_1.

Each pair of driven wheels Z drive two sleeve cranks, one to front bank, one to rear bank.

Front cover scavenge pump drive

Sleeve

Ball socket

Rear bank conn. rods

Front bank conn. rods

Crank

Driven wheels Z

Common Port

Induction

Common Port

Exhaust

Intermediate wheel & pinion Y

Bevel wheel on spider

Driven wheel X

Fixed bevel

Spider shaft (airscrew drive shaft)

Driving bevel

9969

FIG. 7-2 Bristol Hercules VI cutaway. The complexity is more apparent than real, since a poppet valve engine needs cam drives, tappets, and many other components that are absent in sleeve-valve engines. (*Courtesy of Rolls-Royce, Limited.*)

FIG. 7-3 Late Bristol Hercules VI, 14 cylinders, 1650 hp. The spark plugs are close together in the recess of the "junk head." *(Courtesy of Rolls-Royce, Limited.)*

used Merlins. The firm continued to build engines but not large ones. Its main wartime product was the 350-hp Cheetah used in the Avro Anson, a craft occupying a place in the British training scheme that was somewhat analagous to our own Cessna "Bamboo Bomber" (Fig. 7-4).

One can only speculate on the decline of Armstrong-Siddeley, but certain ideas seem reasonable. As a unit of a large steel and armaments combine, a "branch plant" type of management may have existed that could not compete against the sort of strong leadership exemplified by Bristol's Fedden.[1] The fact is that the airframe division of

---

[1]Fedden's dynamic leadership undoubtedly "made" Bristol's engine division, but it always made Bristol's directors uncomfortable. The government knighted Fedden in 1942, not only in recognition of his contributions to British aviation and the war effort, but also to strengthen his hand in his dealings with his trolley-minded board. It was a miscalculation. The Bristol board feared that Fedden's knighthood was a prelude to his somehow displacing them, and so they fired him.

the firm likewise declined from a dominant position to a minor one during the same period. In any case, during the mid-thirties the British big-engine picture was shaking down to a simple pattern of Merlins and a few radials, and, apparently, in every case the Bristols were found to be superior to the corresponding Armstrong-Siddeleys.

In view of the developing shakeout in British large radials, Alvis's ambitious entry into the field in 1936 is incomprehensible unless one is familiar with the firm's past history. Alvis was a small Coventry automobile builder. Under the long-term direction of Mr. T. G. John, the firm had been in and out of trouble at least twice because of impulsive major policy changes. Now, in the euphoria of one of its prosperous periods, Alvis contracted with Gnome-Rhone to build the K series Gnomes in Britain. The firm raised new capital, built a biggish factory, and announced an extensive line of engines, all with Greek names. What makes this interesting in context is the fact that, until just a few years before, Gnome had been a builder of Bristol designs. It can be safely assumed that much of Gnome's know-how had come from building Jupiters; now this knowledge was to return to Britain as competition for Bristol.

The Alvis line included, on paper at least, the 18-cylinder Alcides of up to 1470 hp, the smaller 18-cylinder Pelides, and the 14-cylinder Maeonides. None of these saw production, but Alvis's luck held. With the approach of the war, the government had Alvis use its modern facility to build Merlins. In fact, several "shadow factories" (roughly the equivalent of branch plants) were eventually placed under Alvis's management.

After learning the facts of life with regard to selling a new line of large radials, Alvis in 1939 added a smaller nine, the Leonides, roughly equivalent to Pratt and Whitney's Wasp Junior. This, in the end, turned out to be the last radial engine built in Britain, production continuing well into the sixties. As in the United States, the 450-hp radial fell into a small but definite niche between the turbines and the large opposed sixes. The Leonides went into the post-war Provost trainer and the 10-passenger Percival Prince/Sea Prince/President, an airplane that is a strong contender for the title of the most beautiful small transport ever built.

FIG. 7-4(*a*) Last of the big Armstrong-Siddeley radials, the Tiger VII (ex "Jaguar Major") of 920 hp. Compare with Fig. 6-6: while the Tiger is far more modern looking at first glance, close examination leaves one with the impression that it is pretty much the same engine updated in detail.

The Two-Row Radial Comes to the United States

In the United States there were no big two-row radials before 1932. The background of the first of these goes back to Pratt and Whitney's introduction, in 1929, of its Wasp Junior. The steady uprating of the original Wasp had taken it out of its original power class, and so, to get back into this important market, a new engine of 985 cu in. (16.18 L) had been developed. Like the Wasp, the Wasp Junior had "square" bore and stroke proportions, both, in this case, being 5.1875 in. (131.76 mm). Introduced at the same 300 hp as the J-6-9 Whirlwind of similar displacement, it was up to 420 hp by 1932, and racing versions had stood up to considerably higher outputs.

Wasp Juniors are still working for a living at the age of 41, spraying fields and orchards in Schweitzer-Grumman Ag-Cats and performing at air shows in Stearmans. They seem to have been proof against poor installations; they gave satisfaction in every aircraft that ever used them. They went into Howards, numerous racers, some models of the Beechcraft Staggerwing, Lockheed and Beechcraft light-twin transports, Vought's great OS2U Kingfisher, the Grumman Goose amphibian, and several World War II basic and advanced trainers. The R-985 was almost certainly the world's best engine in its class.

In 1932, Pratt and Whitney used 14 Wasp Junior cylinders in its new Twin Wasp Junior, better known as the R-1535 (for its displacement). It gave 625 hp at its introduction and 725 the following year. A comparison with the Hornet gives a classic example of the relative advantages and disadvantages of a big nine in relation to a fourteen. By 1934 the same power could be had from either engine, at a weight of 880 lb (400 kg) for the Hornet as against 1162 lb (528 kg) for the Twin Wasp Junior, but the Hornet's diameter was 55 7/16 in. (1408 mm) and the R-1535's was 43 ⅞ in. (1114 mm). It is thus clear that drag would be

FIG. 7-4(*b*) Armstrong-Siddeley Cheetah, 350 hp. An excellent engine but somehow less impressive than the Jacobs (Fig. 8-12) that constituted its nearest American counterpart. (*Courtesy of Rolls-Royce Limited.*)

less and top speed higher in an aircraft designed around the two-row engine but that the lighter weight of the nine could be expected to result in better climb and a shorter takeoff run. The fact is that American experience with both types in military aircraft generally showed better results with the Hornet and the comparable Cyclone than with their two-row counterparts. As long as the needed power could be had from 9 cylinders, two-row radials occupied a secondary position in the United States.

The aircraft that used the R-1535 hold considerable interest, representing the transition then going on from the older biplane types to the encroaching low-wing retractables. Grumman's first fighter biplane, for instance, combined wire-braced biplane wings with retractable gear and a cockpit canopy. As the F2F-1 and the F3F-1, it used the R-1535, went 231 mph (372 km/hr), and was not at all an unreasonable-looking airplane, all things considered; as the F3F-3, with a Cyclone, it was hideous but took off better from carriers. Figures available on the change, if any, in top speed are conflicting, but there was no significant loss (Fig. 7-5).

The R-1535 also powered the Vought SB2U Vindicator, the Navy's first low-wing retractable-gear dive-bomber, a plane that was a precursor of future design in contrast with the way the F3F was a modernized hangover from older ideas. Also, it was used in the breathtakingly beautiful 352-mph (567-km/hr) Hughes Racer, now in the NASM. There is no question that the Twin Wasp Junior attracted considerable attention from airframe designers, but, in retrospect, there is room to question whether it really led directly to any better airplanes than could have been built with a Hornet or a Cyclone.

And Pratt and Whitney itself was not completely happy with the R-1535. For one thing, there were un-Pratt and Whitney-like reliability problems, chiefly with the center main bearing. These troubles were cured, but there was another difficulty. The firm decided that it was trying to move in too many directions at once and that its engineering leadership was spread thin, in contrast to what was happening at Curtiss-Wright, where a very tight act had been put together (see Chap. 6). By 1937, the

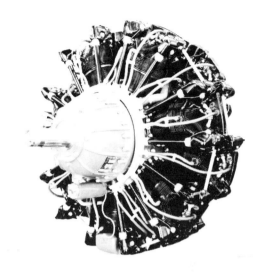

FIG. 7-5   Pratt and Whitney R-1535 Twin Wasp Junior. This is a relatively late SB4-G model of 1934, developing 825 hp. Note that in contrast to the horizontal cooling fins on the heads of earlier radials, those on the R-1535 are vertical. Most high-powered radials after 1930 were made this way. *(Courtesy of Pratt and Whitney Division, UTC Corporation.)*

East Hartford firm was building the Wasp Junior, two types of Wasps, the F and G Hornets, the R-1535 and the R-1830 Twin Wasp, as well as doing preliminary investigation into some entirely new ideas at Army Air Force instigation. Something had to go, and the company decided, correctly as it turned out, that it was the Twin Wasp Junior that should be deemphasized.

This was the right decision because the future of the two-row engine lay not with compact alternatives to big nines but with outputs that were too high to be obtained from good nines, although the R-1830, the firm's mainstay for the next few years, was on the border line between these two categories. The R-1830, named the Twin Wasp (though nobody ever called it that), came out in 1932 with a choice of 6 to 1 compression and 775 hp or 6.5 to 1 and 825 hp. In both cases it developed its rated power at 2400 rpm. This was somewhat fast for a 30-L engine at the time but proved to be well within its capabilities; by 1936 it was giving 1000 hp at 2600 rpm (Fig. 7-6).

A comparison with other 1000-hp radial engines of the same period shows the Pratt and Whitney touch (keeping in mind that whatever Pratt and Whitney could do, Wright could pretty well match). Hispano-Suiza's Wright-based 14A required 2816 cu in. (46 L), 54 percent more than the Twin Wasp, to deliver its 1000 hp at 2100 rpm and, despite what would seem to be a more conservative design, was not nearly as reliable in service as the R-1830 was. Gnome-Rhone's K14 Mistral Major displaced 2360 cu in. (38.7 L) for an output of 1025 hp. The Piaggio P XI R/C 30 needed 2293 cu in. (37.6 L) to produce 950 hp. On the other hand, Bristol, while it had nothing quite up to the 1000-hp class yet, was getting comparable specific output from both the Pegasus and the Perseus. Armstrong-Siddeley, resting on its oars while its smaller engines sold well, could do no better than 880 hp from 1996 cu in. (32.7 L). But, in New Jersey, Curtiss-Wright had gone to about the limit for a nine with the G series R-1820 Cyclone, matching the R-1830's takeoff rating and doing it at 400 rpm less—though at some cost in reduced smoothness. DC-3 buyers had the choice of Cyclone or Twin Wasp power, and it was said that pilots from American Airlines, which used the R-1820, could be identified by their tremor, which took several hours to go away after a long flight with a pair of Cyclones. A gross libel, undoubtedly.

The original Wasp lost much of its market by 1936. In 1932, Boeing was using it, at about 550 hp, in the P-26 fighter, in the B-9 bomber, and in the 247 transport—all advanced aircraft for their day. But the next generation of planes in each category—Seversky's and Curtiss's fighters, Martin's B-10 bomber, and Douglas's DC-2 transport—all used Cyclones. The R-1830 was available in the same power class, but Pratt and Whitney first had to overcome the setback that had resulted from the Boeing 247 imbroglio.

The 247 had come out at exactly the right time. Just when the notorious TWA Fokker crash that killed Knute Rockne was shown to have been caused by rot in the wooden wing structure, here was a clean, fast, strong, all-metal replacement available. Unfortunately, it wasn't really available. Boeing, a part of United Aircraft, would be happy to sell 247s to all comers as soon as United Air Lines' (also a part of United Aircraft) needs were met—a matter of about 2 years, during which United's competition would be stuck with Fokkers and Fords. Jack Frye of TWA refused to hold still for this, and the result was the DC-2. Powered by Cyclones instead of Wasps, seating 18 instead of 10, and having full headroom in the cabin instead of having a wing spar over which the passengers had to step, the DC-2 obsoleted the 247 the day it first flew. Had the 247 been the larger Hornet-powered plane that Boeing wanted to build,[2] it might have been different. As it was, the residue of airline resentment added to the too-small size of the 247 undoubtedly created problems for Pratt and Whitney until, in 1934, the Air Mail Act broke up the vertical integration that had proven to be both a strength and a weakness for United Aircraft.

The R-1830 was a classic engine but did require the highest octane gasoline available. For example, the -82 of 1939 developed 1000 hp but demanded 100 octane. This was a drawback to the airlines from the standpoints of plug fouling and fuel cost. Accordingly, when Douglas proposed the DC-4 as a sort of double-size DC-3, Pratt and Whitney agreed to build an enlarged Twin Wasp, the R-2180, to give the same power as the R-1830 but on 87 octane. This DC-4 was a failure. It was sold to Japan, retroactively named the DC-4E, and replaced by a new design with higher wing loading and more speed, again called DC-4. For this application, the East Hartford firm again obliged with the R-2000. The new DC-4 came out too late to see much civilian service before the war but was built in quantity for the Army as the C-54 and for the Navy as the R5D. A respectable number of R-2000s thus had to be manufactured. The price of 100 octane soon came down to little more than that of 90 octane, so it might be considered that the R-2000 was an American example of solving a nonproblem. However, modification of the R-2000 to use 100 octane provided a useful boost in power from the original 1250 hp to 1450 in the later versions.

---

[2]It is claimed that United Air Lines pilots dictated a maximum size beyond which a fast twin-engine transport couldn't be handled safely.

Pratt and Whitney's next step up in size was to 2800 cu in. (45.9 L) with America's first 18-cylinder radial, the Double Wasp. Much smaller than the world's only other modern eighteen, the Gnome-Rhone 18L of 3442 cu in. (56.4 L), it was nevertheless more powerful, and heat dissipation was correspondingly more of a problem. This meant that, for the R-2800, the cast or forged cooling fins that had served so well in the past had to be discarded. The cooling fins needed were so thin and fine-pitched that they had to be machined from the solid metal of the head forging. I once had a chance to see this operation. All the fins were cut together. A gang of milling saws was automatically guided as it fed across the head so that the bottom of the grooves rose and fell to make the roots of the fins follow the contour of the head. It was a case of designing an engine component that could only be made by a new method and then keeping everything crossed until the new method proved to be practical. In addition to the new head design, the Double Wasp had probably the most scientific baffling yet to direct the flow of cooling air, more so even than the excellent arrangements on the Ranger inline air-cooled engines (Fig. 7-7).

The results were worth the trouble: 2000 hp was obtained from 2800 cu in. (45.9 L), or 1 hp/1.4 cu in. (43.6 hp/L) of displacement. In 1939, when the R-2800 was introduced, no other air-cooled engine came close to this figure, and even liquid-cooled ones barely matched it. The designing of conventional air-cooled radial engines had become so scientific and systematic by 1939 that the Double Wasp was introduced at a power rating that was not amenable to anything like the developmental power increases that had been common with earlier engines. It went to 2100 hp in 1941 and to 2400 late in the war, but that was all for production models. Experimental models, as always, were coaxed into giving more power, one fan-cooled subtype producing 2800 hp, but in general the R-2800 was a rather fully developed powerplant right from the beginning.

This big, hairy engine never went into a heavy bomber. It was exclusively a powerplant for fighters and medium bombers during the war, being used in the P-47, the F6F Hellcat, and the F4U Corsair, and also in the B-26 and A-26 twin-engine mediums. Post-war its reliability commended its use for long-range patrol planes and for the DC-6, Constellation, Martin 4-0-4, and Convair transports. This last application is noteworthy, since these were twin-engine craft of size, passenger capacity, and high wing loading comparable with the DC-4 and the first Constellations. I, for one, had my doubts at the time. Two engines were all right for transports with the DC-3's moderate wing loading, and the high wing loading of the DC-4 was safe enough when there were four engines, but all that weight with only two engines seemed like tempting fate. However, the Convair engineers knew what they were doing. (Those at Martin, and those who tested the Martin for government approval, didn't; the Martin's wings failed from fatigue after a while.) The Convairs were just as good in their way as the four-engine transports. A well-engineered installation and good controls were probably what made the difference; had the Convair encountered as many engine failures as did Curtiss-Wright's deplorable C-46 Commando, it would have been a sad story.

The demands of a major war tend to pull a large builder of military hardware in two different directions. There is constant need for improvement and no problem about getting money, but there is also a continuing demand for production of existing products if they are any good at all—and sometimes, as with the P-40, when they aren't all that useful. In Pratt and Whitney's case, this conflict caused the company to overleap the whole 3000-cu-in. (50-L) class, leaving it to its vigorous New Jersey competitors, and to jump to the 4000-cu-in. (65-L) category.

Thus Pratt and Whitney's next, and, as it proved, last, big piston engine was the Wasp Major, or R-4360, of 71.5-L capacity. This was a four-row 28-cylinder engine, introduced at 3000 hp. Its date of introduction is somewhat obscure. It saw service only at the very end of the war, in late B-29s that were actually early B-50s, but the B-36, as early as 1941, seems to have been planned around the R-4360. Probably there was so much talk and rumor about forthcoming monster engines that Convair was confident that *some* suitable engine would be available in ample time for its proposed giant.

Not too many years earlier, cooling the rear

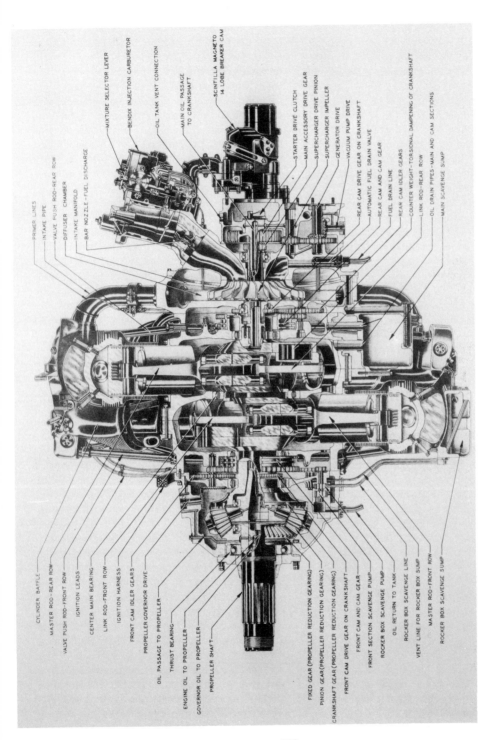

Labels (top group, reading along the pointer lines):

PRIMER LINES
INTAKE PIPE
VALVE PUSH ROD—REAR ROW
DIFFUSER CHAMBER
INTAKE MANIFOLD
BAR NOZZLE—FUEL DISCHARGE

MIXTURE SELECTOR LEVER
BENDIX INJECTION CARBURETOR
OIL TANK VENT CONNECTION

MAIN OIL PASSAGE TO CRANKSHAFT
SCINTILLA MAGNETO
14 LOBE BREAKER CAM

STARTER DRIVE CLUTCH
MAIN ACCESSORY DRIVE GEAR
SUPERCHARGER DRIVE PINION
SUPERCHARGER IMPELLER
GENERATOR DRIVE
VACUUM PUMP DRIVE

REAR CAM DRIVE GEAR ON CRANKSHAFT
AUTOMATIC FUEL DRAIN VALVE
REAR CAM AND CAM GEAR
FUEL DRAIN LINE
REAR CAM IDLER GEARS
COUNTER WEIGHT—TORSIONAL DAMPENING OF CRANKSHAFT
LINK ROD—REAR ROW
OIL DRAIN PIPES—MAIN AND CAM SECTIONS
MAIN SCAVENGE SUMP

Labels (left group):

CYLINDER BAFFLE
MASTER ROD—REAR ROW
VALVE PUSH ROD—FRONT ROW
IGNITION LEADS
CENTER MAIN BEARING
LINK ROD—FRONT ROW
IGNITION HARNESS
FRONT CAM IDLER GEARS
PROPELLER GOVERNOR DRIVE

Labels (bottom-left group):

OIL PASSAGE TO PROPELLER
THRUST BEARING
ENGINE OIL TO PROPELLER
GOVERNOR OIL TO PROPELLER
PROPELLER SHAFT

FIXED GEAR (PROPELLER REDUCTION GEARING)
PINION GEAR (PROPELLER REDUCTION GEARING)
CRANKSHAFT GEAR (PROPELLER REDUCTION GEARING)
FRONT CAM DRIVE GEAR ON CRANKSHAFT
FRONT CAM AND CAM GEAR
FRONT SECTION SCAVENGE PUMP
ROCKER BOX SCAVENGE PUMP
OIL RETURN TO TANK
ROCKER BOX SCAVENGE LINE
VENT LINE FOR ROCKER BOX SUMP
MASTER ROD—FRONT ROW
ROCKER BOX SCAVENGE SUMP

FIG. 7-6(a) Cross section of Pratt and Whitney R-1830 Twin Wasp, one of the best all-around radial engines ever built. Interesting details abound. The crankshaft can now be seen to be in one piece and the master rods are the split type, unlike the original Wasp. The narrowness of the center main bearing is astonishing. Cams and cam drives had to be provided at both ends of the engine because of the two-row configuration. Note that the reduction gear is what a few years earlier was called the "Farman type." The spark plugs are as nearly vertical as it is possible to place them. The pistons show internal ribbing and finning to strengthen the piston heads and to transfer as much heat to the oil as possible. Note also the injection carburetor. Fuel is sprayed under pressure at a point downstream of the venturi; the venturi is used only for sensors to control fuel delivery.

122

FIG. 7-6(b) Pratt and Whitney R-1830 Twin Wasp, 1200 hp, 1941. Used in the Grumman F4F-4 Wildcat, it had what was probably the world's first production two-stage supercharger. When Wildcat production was phased out at Grumman, to allow concentration on the F6F, General Motors built Wildcats with Cyclone power. Top speed (and appearance) suffered but takeoff improved, and this was an entirely appropriate change, since with Hellcats replacing Wildcats in first-line service, the GM version was used on short-deck escort carriers. *(Courtesy of Pratt and Whitney Division, UTC Corporation.)*

FIG. 7-7 Pratt and Whitney R-2800 Double Wasp, 2000 hp, 1942. *(Courtesy of Pratt and Whitney Division, UTC Corporation.)*

cylinders of a four-row engine would have loomed as a daunting task, but Pratt and Whitney made it look easy. Each row of 7 cylinders was staggered behind the one in front of it at a relatively small angle, making each bank of four look like a canted inline. The back side of each bank was closely baffled, the front not at all. So well did this work that it was possible to place each of the seven magnetos in front of the bank it served, despite the effect that this must have had on cooling airflow (Fig. 7-8).

Despite its size and complexity the Wasp Major was a reliable engine in flight, but it was troublesome on the ground. It was one of the (fortunately) few engines ever built that could be shut down in perfect operating condition and have something wrong with it the next time it was started up. In the B-36, the Army Air Force developed a

FIG. 7-8 Pratt and Whitney R-4360-4 Wasp Major, 3000 hp, 1945. Probably the most astonishing thing about this engine is that the air passing through the space between the front cylinders was adequate to cool the 21 cylinders in the remaining rows. At least, it was in flight. The R-4360 did tend to cook itself on the ground, although perhaps not as badly as the R-3350. Note that the rocker arms are no longer in forward-facing pairs. Instead, the intake port faces forward at approximately 45° to the engine's centerline and the exhaust is 180° from it, diagonally to the rear. In 1930 the Franklin automobile firm had found that the most effective cooling for the smallest amount of air was obtained if the air passed over the intake port before cooling the exhaust, but the R-4360 was the first aircraft engine to incorporate this feature in its design. *(Courtesy of Pratt and Whitney Division, UTC Corporation.)*

series of checks so extensive that the shutdown procedure took 6 hours and the start-up 4 hours. It was the R-4360 that gave the Bendix Engine Analyzer the opportunity to prove its usefulness. This device used a cathode-ray tube to make checks on almost every aspect of an engine's operation while it was running. Tried on the B-36, it showed the way to simplifying the horrendous shutdown procedure. Unfortunately, its installation was extremely complex, requiring a sensor and corresponding wire for every operating characteristic that it monitored. For instance, each spark plug had to be connected to the analyzer, making 56 wires per engine for this purpose alone. Under the circumstances, it is hardly surprising that the engine analyzer did not become standard equipment on the six-engine B-36.

The B-36 was one of the very few completely successful shaft-drive airplanes. The big R-4360s were buried in the thickest part of the wing and drove the pusher propellers through long rearward extension shafts. Probably the main reason for its success was the large number of cylinders. A four-cycle engine with 28 cylinders has 14 power impulses per revolution, reducing the torque variation to a negligible factor in comparison with the 6 impulses that, to use a common example, a twelve would produce.

In its other installations the Wasp Major drove its propellers in the conventional manner. These were the B-50 and Martin's Mercator and big single-engine Mauler. Civil applications were few, only Boeing's B-50-derived Stratocruiser achieving significant production. In Northwest Airlines' Stratocruisers a big-engine installation problem that had earlier bothered the British but not ourselves emerged: propellers began to give trouble. In Northwest Orient's Stratocruisers, the Curtiss Electric propellers that Northwest specified would on occasion shift to low pitch and refuse to feather. Blades were thrown, and operation with one engine out was a source of at least one ditching in the Pacific. Fortunately, the Hamilton-Standard hydraulic propellers used by most other Stratocruiser operators were less trouble-prone.

Wasp Majors were also used on the big Douglas Globemaster, the military transport that marked the swan song of the big reciprocating engine. The turboprop-powered Lockheed C-130 Hercules came out while the Globemaster was still in production, and its ability to carry larger loads faster while yet being a less ponderous aircraft was all the demonstration the military needed. After 1958 the end was in sight for the wonderful monsters that had powered the American planes, which had dominated the post-war world's transport market.

Pratt and Whitney was ready. By the late fifties the only radials still being made at East Hartford were the R-2800 and the R-4360. The old faithful Wasp, the perdurable Wasp Junior, and the Hornet were gone. After 30 years, the firm was preparing to bow out of the radial engine game to which it had contributed so much and which had made its name known all over the world. It would go on to perhaps greater things, but something unique was gone.

Curtiss-Wright, by then universally referred to as plain "Wright," was the third member of the triumvirate of major radial engine builders. In point of total horsepower produced and sold, it was in fact number one. Wright came to believe in basing its designs on a larger cylinder than Pratt and Whitney liked. Up to the early thirties the reverse, if anything, had been true. The original Cyclone had used 194.4-cu-in. (3.19-L) cylinders, for a total displacement of 1750 cu in. (28.7 L), as against the early Hornet's 207-cu-in. (3.39-L) cylinder and 1864-cu-in. (30.5-L) total. After about 1932, however, the two firms went in opposite directions. The A series Hornet exemplified the idea of getting at least the same total power from smaller cylinders, running at higher rpm as necessary. Its displacement dropped to 1690 cu in. (27.7 L), or 188 cu in. (3.08 L) per cylinder. A year or so later, by contrast, the Cyclone was enlarged to 1820 cu in. (29.8 L). The enlarged F series had become, by 1935, somewhat overdeveloped in that its maximum horsepower had to be used carefully if overheating and other problems were to be avoided. It will be remembered that the early Hornet also had its problems, and, just as with the Hornet, a very fine engine emerged when these problems were cured. In the case of the Cyclone, the G series got more cooling fin area, a stronger master rod, and a general structural stiffening-up. With these changes, the R-1820 coped very well with outputs that eventually reached 133 hp per cylinder. It would appear at first glance that Wright made the better move. The new, bigger Cyclone became a very widely used engine while the revised Hornet rather faded from the market. However, it should be remembered that the lessons learned about designing a big engine to run at high speeds served Pratt and Whitney well over the next 18 years (Fig. 7-9).

At the same time that it enlarged the Cyclone, in 1933, Wright took out insurance against this being the wrong move by building a small-diameter two-row engine. This was the largely forgotten R-1510, unimaginatively named the Whirlwind. This 14-cylinder engine, obviously very much the counterpart of the R-1535, was apparently never used

in a production airplane. Its peculiar destiny was to power Curtiss-Wright prototype airframes in competition with other engines that were invariably selected in its stead. Thus the Goshawk, the SBC Helldiver biplane (second of the three Curtiss dive-bombers to bear that name), and the second, twin-engine, Shrike were all tested with both the R-1510 and the Cyclone, and in each case the R-1820 was selected as the production engine. In addition, a few 900-hp R-1670s were built, meeting the same fate. Two of these were tried in an A-18 Shrike, and the first XP-36 had one. In the XP-36 it disgraced itself by breaking down before the test program could be completed, and Curtiss-Wright had to suffer the ignominy of seeing a competitor's engine, the R-1830, selected as the powerplant for the production version of the P-36. The fact was that Wright's early two-row engines did not inspire much confidence even at home. The firm selected the bulkier, but lighter and simpler, Cyclone to power the fixed-gear Hawk 75 export version of the P-36 that went to China, Siam (now Thailand), Argentina, and Norway.

After its disappointing experience with the R-1510 and the R-1670, Wright stayed with the big cylinders. Thus its next move was to the R-2600, announced in 1939 at 1500 hp. Like the R-2800 a year later, this 43-L engine was mature enough from the beginning to have relatively little potential for further development. It went to 1600 later in 1939 and to 1800 during the war, but that was all that Wright cared to extract from it. The days of 50 percent power increases during the production life of an engine were about over, at least for air-cooled engines (Fig. 7-10).

The R-2600 made some things practical that wouldn't have been practical in the absence of a good big engine of its size, and the first of these was transatlantic passenger travel. Boeing's big model 314 Clipper for Pan American used four of them. It is hard today to appreciate the impact that those big ocean-spanning flying boats made on the world. France, with the Latècoère 521, had tried to design such a boat a few years earlier but had to employ six water-cooled Hispano 12Ybrs engines to get the size and performance it needed. The Latè was flown to the United States, but in the end all it did was demonstrate that six engines were not

FIG. 7-9 Wright R-1820 Cyclone, 1525 hp, 1953. The basic R-1820 stayed in production for over 25 years. *(Courtesy of NASM-Smithsonian.)*

said for the Empire: considering that every significant British flying boat up to that time had been a an old-style biplane, Short did a terrific job of designing a clean cantilever monoplane in its first effort in this direction.

Another point, one which has little to do with engines but which may be of interest, concerns the fact that each of Pan Am's big boats came from a different maker. The fact was that Pan Am's Juan Trippe was one of the tightest-fisted New Englanders that ever came down the pike, and nobody ever made a dime out of building airplanes to his special requirements. This didn't bother Igor Sikorsky too much; fine man that he was, his main concern was keeping his beloved employees working.[3] Glenn L. Martin, on the other hand, was more like Trippe than he was like Sikorsky, and when Pan Am let him down, as he saw it, by not giving him enough follow-on orders for the 130, he saw the light and got very cautious in his dealings with

the answer. Sikorsky had built its nice S-42 with four 750-hp Hornets, and Martin the model 130 China Clipper with four 830-hp R-1830s, but the Sikorsky was economical only for Caribbean distances, and the Martin for the short-stage Pacific route (Fig. 7-11).

Pan American was not allowed to start transatlantic service until BOAC received, from Short Brothers, equipment that allowed it to operate a counterpart service. These Short Empire boats again serve to illustrate the advancement attributable to the availability of the R-2600. They were powered by four Bristol Pegasus radials of 910 hp each. To this day the British seem to consider the Empire and its military counterpart, the Sunderland, as almost an all-time high point in flying-boat design, but the fact is that their performance was barely in advance of the Sikorsky S-42 of several years earlier and was inferior to that of the Martin. Had not BOAC been heavily subsidized, there could have been no thought of using such uneconomical equipment. One thing, though, must be

FIG. 7-10 Wright R-2600, 1600 hp. Despite the picture's limitations, the very fine-pitched finning of the heads is noticeable. *(Courtesy of the author, taken at the Glenn Curtiss Museum of Local History, Hammondsport, New York.)*

[3]Rentschler put Commander Wilson, by then a United Aircraft executive, in charge at Sikorsky aircraft to try to get things on a more businesslike basis about the time the Pan Am order was in the shop.

FIG. 7-11 The Boeing model 314 Clipper, four Wright R-2600s. The first aircraft to fly paying passengers across the North Atlantic. *(Courtesy of NASM-Smithsonian.)*

Trippe. Boeing had less to fear from Trippe's hard bargaining because it had an unusual aptitude for making development expenditures for one aircraft pay off in the design of another. Thus the 247 made good money using much of the technology developed for the less successful B-9, the B-17 had an offspring in the Stratoliner, and the B-29 gave birth to the Stratocruiser. In the Clipper's case, 314 wings and power eggs, slightly modified, came from the then giant XB-15.

The R-2600's greatest fame came from some of its applications in planes smaller than the big Boeings. Best known of these was probably the Grumman TBF Avenger, one of the relatively few aircraft that gave satisfaction throughout its production life without requiring substantial power increases over the original versions. The Navy's torpedo bomber situation in 1940 was poor (though not nearly as bad as that of the Fleet Air Arm of the RAF). Its only torpedo plane was the Douglas TBD Devastator, which, powered by an early R-1830 of 850 hp, went only 206 mph (332 km/hr) and could not carry an adequate weight of defensive armament. The Navy might have followed the lead of the Italians and sacrificed defensive armament for performance, but the coming of the R-2600 let the Navy have its cake and eat it too: the TBF combined a top speed of 270 mph (435 km/hr) with both upper and lower defensive guns. The same

engine powered the TBF's dive-bomber contemporary, the Curtiss SB2C Helldiver, a worse airplane but a better weapon than the Douglas SBD Dauntless that it replaced. Had Brewster had more resources and fewer Nazi sympathizers among its work force, the Brewster Buccaneer, also using the R-2600, might have been the SBD's replacement.

The R-2600 also went into the B-25, the A-20, the RAF's Martin Baltimore, and the Martin PBM Mariner patrol bomber. This last plane illustrates the importance of good installation engineering. The same engine that was entirely reliable in other installations was somewhat undependable in the PBM. Bermuda Triangle mystery mongers make much of the disappearance of one Mariner on a search mission between Florida and Bermuda, but the facts are that this plane's "squawk sheet" had been written up for smells of leaking gasoline on more than one occasion and that in-flight fires were far from rare on PBMs. In any case, the R-2600 had a somewhat short career after the war; it was relegated to a secondary position, for both military and civilian applications, by larger radials.

With the R-2600 it became clear that engine diameter had ceased to be the determinant of frontal area for radial-powered planes. Some big radials needed air scoops that added nearly 50 percent to the engine's installed height. This was because the engines needed considerable air for purposes other than the cooling of the cylinders. Chief of these needs were those for oil cooling and the supercharger intercooler or aftercooler. When air is compressed, by a supercharger or otherwise, it gets hot. This does little harm when only a moderate degree of supercharge is used, but when a two-stage supercharger is used, it becomes vital to cool the air after it leaves the first stage if the second is to be effective. In addition, large engines have impressive requirements for combustion air. A 2000-hp engine may consume 14,000 lb (6342 kg) of air per hour at full power; this is over 9000 cu ft/min (300 cu m/min) at sea level. In all large-engined planes there was far more air scoop area than would fit the engine's diameter, though it was not always noticeable; for instance, the air intakes for the oil cooler and the supercharger intercooler were in the wing roots on the F4U Corsair and the Curtiss XF15 (Fig. 7-12).

In 1939 the nonviable R-1510 disappeared from the Wright list, and the Duplex Cyclone of 2000 hp, described as "the world's most powerful aircraft engine," was announced. With 18 cylinders and 3600 cu in. (58.9 L), this would have been a formidable powerplant indeed, had it been actually produced, and there is some question as to whether the airframe industry was ready for such a giant. As it was, nothing more was heard of it, and what actually emerged, 2 years later, was the R-3350 (of 55 L), certainly ambitiously large enough. Again, a jump in engine size made some extraordinary flying boats possible. One was the Boeing XPBB1 Sea Ranger, employing two R-3350s in a B-29 wing of 140 ft (42.67 m). Adequacy of other patrol boats and demand for B-29s interdicted further production after the prototype flew, and the Sea Ranger became known as the Lone Ranger.

The other big boat was the Martin Mars. In 1942 this was the world's largest airplane, weighing 72 t (65,000 kg) and spanning 200 ft (60.96 m). In comparison, the next biggest, the contemporary German Blohm und Voss Wiking, weighed 54 t (49,100 kg), had a span of 150 ft (45.7 m), and had to employ six 1000-hp Junkers Diesels.

But the major application for the R-3350 was in the B-29. Since over 4000 B-29s were produced, it is apparent that a tremendous production effort went into these very large engines. In the B-29

FIG. 7-12 Curtiss XF15C-1, Pratt and Whitney R-2800 radial plus jet in tail—a follow-on design to the similar Ryan Fireball. Air scoop under the engine proper plus air intakes in the wing roots. Compare with Kawanishi "Rex," Fig. 1-4. (*Courtesy of the author, taken at Bradley Air Museum, Windsor Locks, Connecticut.*)

there were heavy losses from engine fires during takeoff. Eventually this problem was solved,[4] but the troubles experienced with the R-3350 were the principal reason that the B-29 never earned the respect and affection that crews gave the B-17. Clearly, something was very wrong when the takeoff was the most perilous part of a bombing mission (Fig. 7-13).

As with the PBM, the problem seems to have been with the installation rather than with the engine per se. Post-war, the same engine gave excellent service when put into the Lockheed L-1049 Super Constellation (not the same plane as the Super C or Super G; Constellation designations can be confusing). The R-3350 also had a remarkable post-war offspring called the Turbo-Compound, in which the exhaust gases drove a big

[4]By expedients like eliminating the runup and making the "mag" check while rolling for takeoff (if one can call that sort of thing a solution).

turbine that was geared to the crankshaft to deliver extra power. The Turbo-Compound's takeoff horsepower rose from 2800 to 3400, and its cruising power to 2920. Fuel consumption was extremely low under good conditions. In September 1946, a Lockheed P2V1 Neptune euphoniously named the Truculent Turtle flew 11,236 mi (18,090 km) behind a pair of Turbo-Compounds, a nonrefueling distance record that stands to this day.

Unfortunately, users could not count on this kind of performance. In TWA Super G Constellations, for instance, actual fuel consumption was above, and takeoff power below, the standards on which loading and range standards were based. TWA pilots' response was to take on an extra 600 gal (2270 L) of "pocket fuel"—gas not shown on the weight-and-balance manifest—for transatlantic flights. The added weight and lessened power meant an end to the kick-in-the-back acceleration that had been normal for piston-plane takeoffs. Super G users got used to jetlike takeoffs before there were any jets, and departures from shortish airports like Rome's Fiumcino could be decidedly thrilling. Eventually this was sorted out with the aid

FIG. 7-13 Wright R-3350 18CB, commercial model—2800 hp. Compare with Fig. 7-8; the resemblances are more noticeable than the differences. By the late 1930s, every good large radial was very much like every other one. (Courtesy of the National Archives.)

of in-flight engine analyzers, but it was hairy for a while. I do not know whether this problem existed with the DC-7, another Turbo-Compound user, but the engine was never regarded with the affection and trust that the DC-6's R-2800 received.

Post-war, Wright did not drop its smaller engines to quite the extent that Pratt and Whitney did. It ultimately got into a peculiar arrangement with AVCO, whereby AVCO built Wright engines at Stratford, Connecticut, for military customers while Wright built for commercial applications. This was probably done to allow Wright to divert some of its resources to jet work and to keep AVCO's new facility going until its promising turbine designs were ready for production. The AVCO people felt that they built better Wright engines than Wright did, because of improvements in such things as cylinder nitriding and better shot peening of valve springs. One may be allowed to wonder whether some Pratt and Whitney ideas on engine building might have drifted down from East Hartford.

One unusual post-war Wright engine was the Cyclone 7, or R-1300. As a 7-cylinder engine putting out power equal to that usually generated by 9- or 14-cylinder radials, the R-1300 could be expected to be a somewhat rough engine. Rough it was, at least in the North American T-28 that formed its chief application, but not quite in the way one would expect. The use of Dynafocal-type engine mounts meant that no great amount of vibration was transmitted to the airframe, but the peculiar design of the T-28's exhaust system created a strong roughness of its own. In the T-28, seven exhaust ports were fed into four exhaust stacks intended to provide "jet augmentation" for better performance and the resulting effect was somewhat reminiscent of a V-twin motorcycle.

In any event, the R-1300 gave the T-28 sluggish performance on takeoff, and ultimately a changeover to the good old R-1820 was made to give this post-war trainer the desired briskness on the runway. Of course, the idea that a primary or even a basic trainer needs the kind of performance provided by 1000 hp is patently ridiculous, but military organizations tend to keep thinking up absolutely essential needs for any given piece of gear until the absurdity of the whole thing becomes inescapable, whereupon everybody goes back to square one. In this case, square one turned out to be the use of Cessna 150s for primary training.

These, then, were the engines that saw Britain and the United States through to the end of the era of big radials. Since literally every large radial in the rest of the world derived from British or American designs, the later history of large radial engines elsewhere is entirely a story of adaptation rather than of origination. Nevertheless it holds considerable interest.

Gaulish Devices

In France the Wright R-1510, license-built by Hispano-Suiza, lasted longer than it did at home. Hisso followed Wright designs very closely except for employing its own design of reduction gear. This was partly its undoing, because in hard wartime service the Hisso radials developed a habit of losing their propeller shafts. Hisso had other troubles. In the Potez 630, Hisso cylinder heads became apathetic and lost their will to live in as little as 10 hr use; in the original LeO 451, before the engine was changed to a Gnome, the engine as a whole lasted even less than that. Is it possible that the high morale of Free French pilots flying with the British was due to their finally having reliable engines to fight behind?

The Bristol-descended Gnomes were much better engines, though of marginal power output by 1939. The various types are easily identified by the series letter. The K series, the first that Gnome developed when the firm broke away from purely Bristol designs, was its longest-lasting line. The K series comprised the 7-cylinder, 515-hp Titan, dropped after 1939; the 9-cylinder, 750-hp Mistral; and the 14-cylinder, 1025-hp Mistral Major. The Mistral Major was designated the K-14 and was licensed for manufacture in Eastern Europe, Italy, and Japan (Fig. 7-14).

The L was an 18-cylinder engine of 1400 hp. It was dropped at the beginning of the war to allow the firm to concentrate on the K series and the remarkable M. While it may seem to have been a mistake to have stopped development on France's most powerful radial in 1939, the fact is that the L was an excessively heavy and bulky engine for its output. It was not really the engine that France needed.

The M was produced only in 14-cylinder form. As a result of being designed *ab initio* to run at high speed, 710 hp was obtained from 1158 cu in. (19 L) at 3030 rpm, with a diameter of only 37.4 in. (950 mm). No one else, ever, anywhere, got so much power from a radial of such small diameter, and it was apparently a reliable—if heavy—engine. It might have been the ideal engine for light twin fighters, but, instead of building something along the lines of Britain's Westland Whirlwind, the French tended to waste its potential on three-place twins of vaguely defined mission capability. There was talk of a five-row version of the M, to be used for buried installations in the wings of big bombers, but nothing had come of this before the fall of France. Since four rows of cylinders were the most ever successfully used on any radial, this would almost certainly not have worked (Fig. 7-15).

The most-used Gnome-Rhone was the 14N, a modernized K with increased cooling fin area and some internal strengthening. Most versions gave about 1100 hp. It powered the Bloch MB-152, France's only radial-engine fighter, and a number of twin-engine light fighter-bombers. After the French surrender, the Germans used 14N engines for one of aviation's more remarkable machines, the Me-323 Gegant. This wooden giant had originally been built as a glider, but it had to be towed into the air by a troika of three Me-110s—clearly a procedure calling for more nerve than brains on the part of all hands. It was therefore converted into the world's largest-ever power glider by installing six Gnome 14N power eggs, cannibalized whole from newly completed Bloch MB-175s. These aircraft were the unwilling stars of one of the war's more gruesome incidents when a number of them, loaded with about 200 soldiers each in an effort to reinforce Rommel's forces, were caught and shot down over the Mediterranean by Allied fighters.

Japanese Radial Engines

Nakajima built Gnome-Rhone designs in Japan. This firm had provided Japan with its first large radials by building Bristols, from the late twenties

FIG. 7-14   Gnome et Rhone 14K Mistral Major, 14 cylinders, 1025 hp, about 1937. *(Courtesy of NASM-Smithsonian.)*

FIG. 7-15   Gnome et Rhone 14M, 14 cylinders, 710 hp, 1939. No other radial of anywhere near its power even came close to the 14M's 37.4-in. (9500-mm) diameter. *(Courtesy of NASM-Smithsonian.)*

onward, as its Kotobuki of up to 830 hp. When Gnome evolved its own designs, Nakajima added a Gnome-Rhone license in 1936, and its 9-cylinder engine became the Hikari of similar power to the Kotobuki. Actually, since the Kotobuki had fully enclosed valve gear, it was more like a Gnome than a Bristol even before the Gnome license was issued (Fig. 7-16).

Before 1936, Japan's military aviation had been almost as heavily oriented toward liquid cooling as Germany's. Its engines were largely license-built Lorraines and BMWs. Japan, with the Gnome license of that year, and with Mitsubishi's Pratt and Whitney arrangement of 1937, saw a complete reversal take place, and Japanese aviation in World War II was very heavily radial-oriented. There can be little doubt that the Gnome-Rhone and Pratt and Whitney licenses were essential to the Japanese war effort. The Japanese engines that were closely related to their Western forebears were the best they had; those that developed well beyond the original models were troublesome.

FIG. 7-16 Nakajima Kotobuki, model 1, Mod. 1, about 1935. While technically excellent, this manufacturer's photo is decidedly lacking in character. The cooling fins on the cylinder heads are interesting. The Japanese are said to have been unable to produce finely pitched cast fins and this photo may bear out the story, but the designers did not make the task at the foundry any easier by sloping the fins on a converging pattern. (Courtesy of the National Archives.)

Japan and Russia would have had an advantage over many countries in an obscure but important area when it came to building others' designs under license. Each major country had its own types of steel and aluminum, and adapting one nation's specifications to what was available in another country could cause problems. An American maker could order Aircraft Quality 4340 from any of several steel mills and be sure of getting the same product from each one. The German license builders would have to pick the DIN (Deutsche Industrie-Normen)[5] steel whose composition came closest to AISI (American Iron and Steel Institute) 4340 and develop their own heat treating and other procedures to allow for whatever differences existed between the two materials. Working from British designs was much worse, since Britain had

[5]German industry standards.

no standardization whatever. British steel was bought by brand name, each brand representing a particular mill's proprietary, more or less secret, analysis. In Japan and Russia, however, the government would simply order a steel mill to make 4340 to the AISI specifications, or to make a German-type steel to the DIN analysis, and that would be the end of the problem. Much the same situation applied to aluminum, to bronze, and, even more so, to bearing metals.

Tracing Japanese radials is extremely complicated. Every one of Japan's later engines had at least four different designations, and descriptions of aircraft use these designations almost at random. For example, the Nakajima Sakae was at once the Sakae-12 through Sakae-31, the Navy Ha-5 through Ha-115 (not consecutively), the Army type 97 and Type 1 Model 2, and, late in the war, the [Ha-35]. This last represents a joint Army-Navy coded system in which the first number gives the number of cylinders, in this case 14, and the second the bore and stroke, here a 5.1-in. (130-mm) bore by a 5.9-in. (150-mm) stroke. Thus it is probably impossible for nonspecialists to ever reach a point where they can mentally connect most Japanese engine nomenclature with the actual engine referred to (Fig. 7-17).

Where the maker's name and designation is given, the task is easier. The Sakae (Prosperity) mentioned, the powerplant of the once-feared Zero, was a Gnome-derived 1710-cu in. (27.97-L) engine built from 1939 to the end of the war. It seems to have been a very good engine. One of its virtues was an ability to run for long periods on an extremely lean mixture. Operated thus, a Zero had the phenomenal ferrying range of over 1000 mi (1610 km).

Nakajima built two other fourteens, neither much good, and an eighteen. The fourteens were the unreliable pre-war Mamoru (Protector) of 1870 hp, possibly owing something to the Pratt and Whitney R-2000, and the Ha-177/Ha-145/[Ha-44] of 1941 and 1942. This was the most powerful 14-cylinder engine ever built, at up to 2400 hp, and, at 2715 cu in. (44.49 L), one of the largest. Apparently it was also one of the most useless.

The eighteen was the Homare (Honor), Japan's most-used engine of the later part of the war.

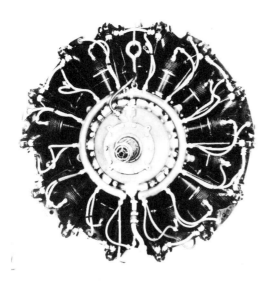

FIG. 7-17 Nakajima Sakae, model 12 (Sakae 12), 14 cylinders, about 1050 hp. No coarse head fins here! Possibly the fins were milled out and the curved section, where the vertical fins turn horizontal, finished by hand. *(Courtesy of the National Archives.)*

It had a fairly short stroke and a comparatively small diameter. Various versions had fan cooling, low-pressure fuel injection, turbosupercharging, and water-methanol injection. All models suffered from unreliability, difficulty in servicing, and vibration. It originally delivered 1850 hp; later versions were rated as high as 2100 but could not be counted on to deliver their rated power, especially at altitude. Among its designations were NK-9A through NK-9S, Ha-39, Ha-109, Army Type 2, and [Ha-45]. Its coding works out to a displacement of 2195 cu in. (35.97 L) (Fig. 7-18).

Mitsubishi built generally better engines, thanks probably to its Pratt and Whitney connections. Smallest was the Zusei (Holy Star) of 1716 cu in. (28.12 L). It was introduced in 1937 at 780 hp and, in Navy versions, gave up to 875. It lasted long enough as an Army engine (type 101) to receive a late-code designation as [Ha-31] and to be made in a 1080-hp version. While reliable, it was too small to see much wartime usage.

The Kinsei (Golden Star) was larger, at 1980 cu in. (32.45 L), and saw service throughout the war. Beginning at 730 hp, it went almost immedi-

ately to 910 hp and gave as much as 1560 at the end. It was designated the Army 99 Model 2, the MK 8-A through MK 8-P, the Ha-112, and finally the [Ha-33].

Largest of Mitsubishi's original trio of 14-cylinder radials was the Kasei (Mars) of 2576 cu in. (42.21 L). It was very widely used, and every imaginable combination of goodies was available in its many marks and models. Despite its large size, the only troublesome subtypes were the Kasei 13, which had an extension shaft, and the late, high-powered (1900-hp) Kasei 23, which used fan cooling and water-methanol injection. The Kasei 23 suffered from severe and destructive vibration at certain critical operating rpm. The Kasei was also called the MK 4-A through MK 4-U, Army type 100, and [Ha-32], as well as a miscellany of various Ha numbers (Fig. 7-19).

Like Nakajima, Mitsubishi ran into trouble when the original inspiration from foreign sources ran out and it had to strike out on its own. The difference was that, in Mitsubishi's case, the outside start was stronger and lasted longer. Its bad engine only came late in the war. It was an 18-cylinder radial of 2546 cu in. (41.72 L) and up to 2200 hp. It carried the designations of Army type 4, Ha-104, Ha-211, MK 9-A through MK 9-C, and [Ha-43]. It was both fan-cooled and turbosupercharged. Unlike the Homare, the MK 9 was little used. Mitsubishi's facilities were too badly needed for the production of Kaseis to be diverted to doubtful projects.

A close examination of the late and desperate efforts made by the Japanese to get reliable engines of adequate power makes one wonder what they thought they were doing when they started the war. With the Army and the Navy each doing its own thing and with no native base in internal engine technology, I can only conclude that there was more hubris than rationality behind the Japanese aggression.

Some of the Japanese engine problems may have been more a matter of poor installations than of basically faulty engines. It is very noticeable that Japanese aircraft were invariably slimmer than their American counterparts. The main reason for this was that, while American designers enlarged their engine cowlings to oval shapes to accommodate

**FIG. 7-18   Nakajima Homare 21, 18 cylinders, 1900 hp, 1943 or 1944. This picture was taken, very much under field conditions, of an engine from a captured aircraft and certainly does not lack character. The head fins may well be *too* closely pitched, and the air resistance caused by turbulent airflow probably caused more air to go over the fins than through them unless the heads had closely fitting baffles that have been removed for the photograph. *(Courtesy of the National Archives.)*

airflow to superchargers, oil coolers, and intercoolers, the Japanese used round cowlings with rather small separate air scoops added outside the cowlings. These tended to look like afterthought engineering. If, in fact, the provision of adequate airflow for hot-weather operation was somewhat neglected, some of the Japanese engine troubles may be traceable to this source. In addition, the Japanese were addicted to the use of fan-assisted cooling. While this can be designed to work well, one wonders whether a system tested in cool Honshū or Hokkaidō would be adequate for the South Pacific or the Philippines.

Germany: The BMW

In Germany there was only one significant line of large radials, the Pratt and Whitney–based BMWs.

BMW began building Hornets in 1930 and quickly overwhelmed its only serious home competitor, Siemens. Actually, Siemens built a pretty good 540-hp radial, its Sh 20B, in the early 1930s, but, despite the use of volute springs to close the valves in an effort to keep the bulk under control, it had the excessive diameter of 57.4 in. (1457 mm). One reason for its size was its low operating speed, 1850 rpm. In the hope of remedying this, the next Siemens, a 750-hp nine, was designed to run at 2330 rpm. Diameter was reduced to 52 in. (1320 mm), but now the valve arrangement was poor. As in some smaller engines, notably certain Italian ones, there was one large rocker-arm cover, and the valves were in line fore and aft; there was no way for adequate amounts of air to get between the valve ports for cooling. Since the Junkers Ju 52-3m was the only aircraft in Germany during the early thirties that used air-cooled radials in the 700-

FIG. 7-19 Mitsubishi Kasei Model 11 (Kasei 11), 14 cylinders, about 1850 hp. Japan's best large engine. Note the difference between the pushrod angle on this engine and that on the Pratt and Whitney types. This resulted from a different cam ring design, one of the major changes made by Mitsubishi from the original. It has been said that the Mitsubishi design was lighter. *(Courtesy of the National Archives.)*

odd-hp class, when the BMW Hornet was adopted for the Ju 52, the large Siemens faded from the market. Siemens changed the name of its engine operation to Bramo, for BRAndenburg MOtoren Werke,[6] and some development went on, but Bramo's main emphasis was on its smaller radials. A Bramo 323R was used in the -Z2 version of the Dornier DO17 bomber early in the war, but that was the last that was heard of the big Siemens/ Bramo. Even with water-methanol injection, no more than a doubtful 1200 hp could be had from the Bramo, and this was no longer enough.

After 10 years of building Hornets, BMW added two two-row radials, largely of its own design, in 1940. The eighteen, the model 802 of 1675 hp, was the less used of the two, though it reached an output of 2400 hp toward the end of

the war. The classic BMW radial was the 14-cylinder 801, powerplant of the formidable Focke-Wulf F-W 190 fighter. It displaced 2562 cu in. (42 L), had a diameter of 48 in. (1219 mm), 52 in. (1320 mm) over the cowling, and normally came as a complete power egg, ready to install in the aircraft. It had a two-speed supercharger which did not give it quite the high-altitude performance that was conferred on the D-B 601 by that engine's fluid-drive blower. A cooling fan permitted a quite small frontal opening for the cowling. Introduced at 1580 hp, it was giving a reliable 2000 by 1945. Altogether, it was an absolutely superb engine. Not the least of its accomplishments was its demonstration that the bulk of a radial had finally ceased to be a handicap in aircraft designed for the highest possible speed.

Italy: Political Engineering?

Italy came late to the use of large radials, and all of its designs can be traced to foreign origins. Most of these were by Bristol out of Gnome-Rhone. It is true that Alfa Romeo had, in 1927, entered the radial engine field at government request, but there was little production of large engines by Alfa before 1935. In that year, Alfa built its 125 R/C, a 750-hp development from the Pegasus. It powered the Cant Z506B Airone (Heron) trimotor bomber that saw some service in Ethiopia. It was thus 10 years after being told to look into radial engines that Alfa engines got their first military application. By 1938, Alfa had mastered the intricacies of cooling an 18-cylinder radial, albeit not a high-powered one, and was producing its 1350-hp Tornado for the Cant Z-1018 Leone (Lion), one of Italy's few twin-engine bombers. The Tornado was based on the Gnome-Rhone L type; like its parent, it did not offer sufficient prospects to encourage further development.

Fiat entered the field in 1933, building the Gnome-Rhone 14 Ksd as its A 58C of 740 hp. The following year, Piaggio also built a Gnome derivative, its 500-hp C. 35[7]; since it was the government rather than any one firm that held the Gnome license, the making of Gnome-based engines was

---

[6]In 1939 or 1940, BMW took over Bramo at the instigation of the government.

[7]Later redesignated P-VII c 35.

not confined to any single firm. In 1934, Fiat took out a Pratt and Whitney license, building the Hornet in the guise of its A 59. As usual with license builders having considerable technical resources of their own, Fiat soon began to make changes. The first evidence of this was the A 78, a Hornet with a Fiat-designed supercharger. By 1937, Fiat had developed its 14-cylinder A 80 R/C 41 of 1000 hp—not reliable in every installation (Fig. 7-20). It worked well in a neglected but significant airplane, the Breda 65. This was a fast single-seat low-wing dive-and-attack bomber, designed when the rest of the world was still thinking of such aircraft in terms of two-seat biplanes. It was the legitimate ancestor of all such large attack planes as the Il-2 Stormovik and the Douglas AD Skyraider series ("Spad" to Vietnam fliers).

Fiat carried the 9-cylinder approach no further, staying with 14 cylinders in its A 74 R/C 38 of 840 hp. This was followed, in 1940 or 1941, by the 1250-hp A 82 R/C 42. Of the two, the A 74 was much the more popular, becoming the mainstay of the Regia Aeronautica, the Italian Air Force (Fig. 7-21).

Piaggio's gradual development of increased independence from its original source of inspiration

FIG. 7-20 Fiat A 80 R/C 41, 14 cylinders, a late engine, made in 1941. Not even Italian engineers could design a 14-cylinder radial to have a neat appearance! (Courtesy of Centro Storico Fiat.)

paralleled Fiat's. By 1935 Piaggio was claiming that its line had diverged so far from its Bristol and Gnome-Rhone origins as to be more its own design than otherwise. In that year, Piaggio listed the 7-cylinder Stella VII C/15 of 440 hp, the 9-cylinder Stella IX R/C 40 of 610 hp, and the 9-cylinder Stella XR of 700 hp. The year 1936 saw the addition of a fourteen, the P XI R/C 40, developed from Stella VII cylinders which gave 910 hp. In 1937 another fourteen, the P XIV R/C 35, apparently a new design, was added. As this engine gave a mere 700 hp, its rationale is far from clear. Perhaps, as discussed below, governmental economic stringency had something to do with it.

Italy fought the war with these engines. Unlike Japan's, Italy's war effort, such as it was, did not run into trouble caused by last-minute efforts to increase power output. Unlike Russia, Italy did not succeed where the original designer had failed in extracting the most possible power from designs. Unlike the Americans, the British, and the Germans, the Italians built no very large aero engines. They simply did a first-class job of building other peoples' designs and of making minor improvements on those designs. There is, of course, this puzzle: Why did they stick to engines of inadequate power? It is well known that Italian fighters were excellent for their power but that their power was inadequate. Italian bombers almost all used three smallish motors instead of two decent-size ones, suffering as a result from the enforced absence of flexible nose armament. Why was this? Why the concentration on engines of marginal size and power?

In part it was because the Regia Aeronautica, like the original Luftwaffe, was built as much as a weapon of bluff as a weapon for actual fighting. In part it was because the Italian military leaders, such as they were under Mussolini, were misled by their easy successes in Africa and Spain. But the main reason may have been economic. Italy was, after all, a relatively small country, yet as late as 1938 it had an air force that was numerically larger than that of the United States. It simply could not afford quantity and quality at the same time. In addition, Italy's economic base had been seriously eroded by the costs of its African empire building and by Mussolini's misdirected drive toward economic self-

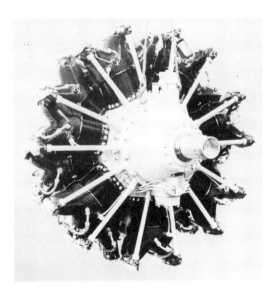

FIG. 7-21   Fiat A 74 R/C 38. While designed earlier than the A 80, its heaviest use seems to have come later. *(Courtesy of Centro Storico Fiat.)*

sufficiency. All in all, the Fascist government had probably bitten off more than it could chew.

Russia and the Post-War Era

Neither in France, in Germany, in Japan, nor in Italy was there any significant post-war history of radial engine development. Only in France were post-war piston-engine transports built, and even these relied on American engines or, in the smaller types, on revivals of pre-war designs. In Soviet Russia, however, there was a long period during which large radials assumed an importance far beyond that of the pre-war era.

If the Fascist countries were Pratt and Whitney territory, the Soviet was largely Wright's manor. Of the Soviet radials, it is known that the ubiquitous M-25 and its derivatives were Cyclone based. The Gnome 14 K appeared in Russia as the M-85, the Bristol Jupiter was made as the M-15; the Gnome modification of the Jupiter was made as the M-22. The 630-hp ASh21 and the 1000-hp ASh62 that constitute the alternative powerplants for the big

AN-2 Colt biplane were apparently Cyclone derivatives, but whether the 1100-hp M-88b of the widely used IL-4 bomber was a Cyclone or a Gnome I have been unable to find out.

The Russian war effort did not rely heavily on radial power. V-12s of Hispano, Curtiss-Wright, and BMW derivation, as explained in Chap. 5, powered most of its military planes. The radial only came into its own in the U.S.S.R. when the Soviets invented air transport after the war. Two large radial engines developed late in the war were used in the IL-2s and IL-3s that formed the backbone of Soviet air transport until the (early) coming of the turbine to Russia: the 1850-hp ASh82 and the ASh73 of 2000 hp. The ASh82 had seen service in the TU-2 intruder bomber, and the ASh73, in the Russian B-29 copy, in the TU-4.

These engines have no personalities that come through to us. When we learn about the ground overheating troubles of the R-3350s in B-29s, or that Bristol engines had automatic boost control to obviate the need for careful monitoring of altitude that characterized many American engines, or that Italian warplanes had chrome-plated cockpit controls, such things bring the dead machinery to life for us: but no information of this sort is available for Soviet engines. About all we know about the big Soviet radials is that they were two-row engines. Travelers who used Aeroflot during the piston era invariably remarked on the casual way that the Russian pilots seemed to just get in and go, with a minimum of preflight or cockpit check, and so we can assume that the Soviet radials had few refinements or complexities; but, again, this is a deduction at best.

Radial engines still have a place in Soviet-zone aviation, but they are no longer produced in Russia. In 1952—probably to free their own resources for turbine production—the Soviets gave the responsibility for supplying radials to Poland. This might seem an odd choice, since both East Germany and Czechoslovakia were more industrialized than Poland before the war. However, Poland, first as Narkiewicz, than as OKL, and finally under the PZL banner, had been doing a good job of developing smaller radials. Thus the old Cyclone-based ASh 62 is now the PZL WSK P-27, and the former Ivchenko A 126B, a big seven of 575 hp, has been the PZL 4T3 since 1963. Though this has been made mainly as a helicopter engine, the government's Pezetel export agency has been promoting it in the West as a replacement for engines that are now out of production. One has been tried out in a Grumman Ag-Cat; apparently it is an engine of impressive quality.

The early design license fees were paid meticulously. It was ideologically correct to do so because it accorded with the Communist dogma that capitalism would contribute to its own demise by being unable to resist selling its secrets. Later the picture changed. The imported designs became more and more Sovietized, and the Russians persuaded themselves that what they got from England and the United States should come to them gratis

because the U.S.S.R. was, in terms of casualties, bearing the brunt of the war. That the whole thing was largely the fault of Stalin's duplicity was, of course, irrelevant.

Now the big radial has almost everywhere gone the way of the coal-burning steam locomotive and is even less likely to make a comeback than that similarly likeable piece of machinery. Few complex manufactured objects have ever reached the perfection of an R-2800 or a Bristol Centaurus. In a first-class installation, such as that in the DC-6, these elaborate machines reached almost unbelievable predictability of operation. But whatever they could do, a turbine can do better, cheaper, and longer. The strong and confident sound of the big round engines is a thing of the past.

Data for Engines

Maker's name	Date	Cyl-inders	Config-uration	Horse-power	Revolutions per minute	Bore and stroke, in. (mm)	Displace-ment, cu in. (L)	Weight, lb (kg)	Remarks
Bristol Perseus	33	9	R, A	580	2400	5.8 (146) × 6.5 (165)	1520 (24.9)	1026 (466)	Perseus XVI gave 955 at 2750 rpm
Bristol Aquila	34	9	R, A	500	3000	5 (127) × 5.4 (136)	950 (15.6)	830 (377)	
Bristol Taurus XII	39	14	R, A	1130	3100	5 (127) × 5.4 (136)	1550 (25.4)	1300 (591)	
Bristol Hercules I	39	14	R, A	1290	2800	5.8 (146) × 6.5 (165)	2364 (38.7)	1680 (764)	Hercules VI gave 1650 hp
Bristol Centaurus	38	18	R, A	2520	2700	5.8 (146) × 7 (178)	3270 (53.6)	2695 (1223)	1938 date is official but unrealistic; not produced until 1942
Bristol Mercury	31	9	R, A	560	2600	5.8 (146) × 6.5 (165)	1520 (24.9)	950 (432)	Destroked Pegasus for fighters
Bristol Pegasus	32	9	R, A	590	2300	5.8 (146) × 7.5 (191)	1753 (28.7)	NA	Late models gave 1000 hp
Alvis Alcides	36	18	R, A	1700	2400	5.8 (146) × 7.1 (180)	3310 (54.2)	2050 (932)	
Alvis Pelides	36	18	R, A	1060	2150	5.8 (146) × 6.5 (165)	2360 (38.7)	1475 (670)	
Alvis Maeonides (Major)	36	14	R, A	700	3000	4.8 (122) × 4.6 (116)	1158 (19)	920 (418)	The only photo of this engine is Fig. 7-15 with an Alvis nameplate retouched in
Alvis Leonides	39	9	R, A	415	3000	4.8 (122) × 4.4 (112)	905 (14.8)	653 (297)	
Pratt and Whitney Wasp Junior	29	9	R, A	450	2300	5.2 (132) × 5.2 (132)	985 (16.1)	653 (297)	Generally referred to as R-985
Pratt and Whitney Twin Wasp Junior	32	14	R, A	625	2400	5.2 (132) × 5.2 (132)	1535 (25.2)	1162 (528)	Late models gave 825 hp
Pratt and Whitney Twin Wasp	32	14	R, A	825	2400	5.5 (140) × 5.5 (140)	1830 (31.6)	1467 (665)	Always referred to as R-1830; late models gave up to 1200 hp
Hispano-Suiza 14A	36	14	R, A	1000	2100	6.1 (156) × 6.7 (170)	2816 (46.1)	1408 (640)	Based on Wright R-2600?

(cont.)

Data for Engines (continued)

Maker's name	Date	Cyl-inders	Config-uration	Horse-power	Revolutions per minute	Bore and stroke, in. (mm)	Displace-ment, cu in. (L)	Weight, lb (kg)	Remarks
Piaggio P XI R/C 30	37	14	R, A	950	2300	5.7 (146) × 6.7 (165)	2293 (37.6)	1366 (621)	Destroked Gnome-Rhone 14K
Curtiss-Wright Cyclone G	37	9	R, A	1000	2000	6.1 (156) × 6.9 (174)	1823 (29.9)	NA	First 1000-hp nine
Pratt and Whitney Twin Hornet A	38	14	R, A	1150	2350	5.8 (146) × 6 (152)	2180 (35.7)	1635 (743)	Better known as R-2180
Pratt and Whitney Twin Wasp D	41	14	R, A	1200	2550	5.8 (146) ×	2000 (32.8)	1585 (720)	Better known as R-2000; used only on DC-4
Pratt and Whitney Double Wasp	39	14	R, A	2000	2400	5.8 (146) × 6 (152)	2800 (45.9)	2350 (1068)	Better known as R-2800; maximum power was 2500 hp
Pratt and Whitney Wasp Major	43	28	R, A	2800	2700	5.8 (146) × 6 (152)	4360 (71.5)	3600 (1636)	Better known as R-4360; maximum power was 4300
Curtiss-Wright R-1510	33	14	R, A	765	2400	5 (127) × 5.5 (140)	1512 (14.8)	1000 (455)	No other data available
Curtiss-Wright R-1670	34	NA	R, A	900	NA	NA	NA	NA	
Curtiss-Wright R-2600	37	14	R, A	1500	2300	6.1 (156) × 6.3 (160)	2603 (42.7)	1950 (885)	Maximum hp was 1800
Curtiss-Wright R-3350	40	18	R, A	2800		6.1 (156) × 6.3 (160)	3350 (54.9)	2779 (1263)	Gave 3700 hp at 1900 rpm in later "Turbo-Compound" form
Curtiss-Wright Cyclone 7	40	7	R, A	800	2600	6.1 (156) × 6.3 (160)	1300 (21.4)	919 (417)	
Gnome-Rhone L-14	36	14	R, A	1000	2150	5.8 (146) × 7.3 (185)	2647 (43.4)	NA	Maximum hp was 1400. Earlier Gnome-Rhone designations put letter first.
Gnome-Rhone M.14	35	14	R, A	810	3030	4.8 (122) × 4.6 (116)	1158 (19)	922 (419)	
Gnome-Rhone 14N	36	14	R, A	1085	2360	5.8 (146) × 6.5 (165)	2360 (38.7)	1300 (591)	Modernized 14 K Mistral Major; better cooling, internal strengthening
Nakajima Kotobuki	35	9	R, A	550	NA	NA	NA	NA	
Nakajima Hikari	36	9	R, A	710	NA	NA	NA	NA	
Nakajima Sakae	NA	14	R, A	1020	2600	5.1 (130) × 5.9 (150)	1700 (27.8)	1175 (533)	Maximum hp was 1360

(cont.)

Nakajima Mamoru	38	14	R, A	1870	NA	NA	NA	NA	
Nakajima Ha-177/Ha-145/[Ha-44]	41	14	R, A	2400		5.5 (140) × 6.3 (160)	2715 (44.5)	NA	
Homare	40	18	R, A	1850	NA	5.1 (130) × 5.9 (150)	2195 (35.9)	NA	
Mitsubishi Zusei	37	14	R, A	865	2540	5.5 (140) × 5.1 (130)	1716 (28.1)	1200 (546)	Maximum hp was 1080
Mitsubishi Kinsei	37	14	R, A	1280	2500	5.5 (140) × 5.9 (150)	1980 (32.5)	1200 (545)	Maximum hp was 1560
Mitsubishi Kasei	37	14	R, A	1300	NA	5.9 (150) × 6.6 (170)	2576 (42.2)	NA	Maximum hp was 1900
Mitsubishi Ha-104/Ha-211/[Ha-43]	42	18	R, A	2200	NA	5.5 (140) × 5.9 (150)	2546 (41.7)	NA	
Siemens Sh 20B	34	9	R, A	540	1850	6.1 (154) × 7.4 (188)	1921 (31.5)	925 (420)	
Siemens SAM 322 H2	34	9	R, A	715	2330	6.1 (154) × 6.3 (160)	1636 (26.8)	1080 (491)	
Siemens 323R	36	9	R, A	1000	2500	6.1 (154) × 6.3 (160)	1636 (26.8)	1199 (545)	Developed from 322; better known as Bramo Fafnir
BMW 802	40	18	R, A	1675	NA	6.2 (156) × 6.2 (156)	3294 (54)	NA	Maximum hp was 2400
BMW 801	40	14	R, A	1580	2700	6.2 (156) × 6.2 (156)	2562 (42)	2702 (1228)	Maximum hp about 2000
Alfa Romeo 125 R/C	35	9	R, A	680	2200	5.8 (146) × 7.5 (190)	1747 (28.6)	1045 (475)	Basically a Bristol Pegasus
Alfa Romeo 135 R/C 34 Tornado	38	18	R, A	1500	NA	5.8 (146) × 6.3 (160)	2940 (48.2)	2100 (955)	Described as original design but very much Bristol in appearance
Fiat A 58C	33	14	R, A	740	2300	5.8 (146) × 6.5 (165)	2360 (38.7)	1157 (526)	An almost pure Gnome-Rhone Mistral Major
Piaggio P-VII c 35	35	7	R, A	500	2100	5.8 (146) × 6.5 (165)	1178 (19.3)	716 (325)	Described as Gnome-derived but looks like a Bristol
Fiat A 80 R/C 41	37	14	R, A	1000	2100	5.5 (140) × 6.5 (165)	2788 (45.7)	NA	
Fiat A 74 R/C 38	NA	14	R, A	870	2520	NA	1907 (31.3)	1257 (571)	
Fiat A 82 R/C 42	40	18	R, A	1250	2100	5.5 (140) × 6.5 (165)	2788 (45.7)	1600 (727)	Based on A-80 cylinders with more cooling fin area
Piaggio Stella VII C/15	35	7	R, A	400	2100	5.8 (146) × 6.5 (165)	1178 (19.3)	NA	Commercial version of P-VII

Data for Engines (continued)

Maker's name	Date	Cyl-inders	Config-uration	Horse-power	Revolutions per minute	Bore and stroke, in. (mm)	Displace-ment, cu in. (L)	Weight, lb (kg)	Remarks
Piaggio Stella IX R/C 40	35	9	R, A	600	2250	5.8 (146) × 6.5 (165)	1515 (24.8)	981 (446)	Looked very much like a Gnome-Rhone K-9; Stella XR was same engine with higher takeoff rating, less supercharge
Piaggio Stella P XI R/C 40	37	14	R, A	1000	2200	5.8 (146) × 6.5 (165)	2354 (38.4)	1444 (656)	
Piaggio Stella P XIV R/C 35	37	14	R, A	700	NA	NA	NA	NA	
Soviet Mikulin M-25	NA	All data apparently same as Wright R-1820							
Soviet Mikulin M-85	NA	All data apparently same as Gnome-Rhone 14K							
Soviet Mikulin M-15	NA	All data apparently same as Bristol Jupiter F							
Soviet Mikulin M-22	NA	All data apparently same as Gnome-built Jupiter							
Soviet Shvetsov ASh 21	NA	9	R, A	630	NA	NA	NA	NA	
Soviet Shvetsov ASh 62	38	9	R, A	1000	2200	6.1 (156) × 6.8 (175)	1818 (29.8)	1184 (538)	Ex-M-62; Cyclone-based
Soviet Shvetsov ASh 82	NA	14	R, A	1675	2400	6 (155) × 6 (155)	2513 (41.2)	1947 (885)	Based on R-1800; maximum hp is 1850
Soviet Shvetsov ASh 73	NA	18	R, A	2000	NA	NA	NA	NA	Said to be based on Wright

R—radial; A—air-cooled; NA—not available. Open figures—U.S. Customary; figures in parentheses—metric. Weights are dry.

EIGHT

The smaller radials.

France: The Anzanis and the Salmsons

The great improvements in engines that took place in World War I were largely confined to powerplants large enough for combat aircraft. Trainers, on the whole, had to make do with less advanced engines. As a result, the development of light airplanes in the immediate post-war period was hampered by the lack of up-to-date small engines.

On the whole, the incentive to develop modern engines was weak at first. While there was a great deal of enthusiasm for aviation in most countries, the cost of flying was high, and the existence of large quantities of war surplus equipment made most manufacturers hesitant to enter the field. This was especially true in the United States, where the combination of JN4D airplanes and OX-5 engines in great numbers and at low prices almost completely eliminated, until perhaps 1925, any chance of survival for manufacturers of airplanes or engines for the private flier.

Things were somewhat more favorable on the other side of the Atlantic. There, most surplus engines were rotaries—ruinously expensive to operate—and there was some government support for private flying.

The French Anzani engines were already pointing to the radial as one of the most promising types for lightplanes. Their small 3-cylinder radials had been the mainstay of the remarkable French training method. This system employed little single-seater airplanes called ''roleurs'' (we called them ''Penguins'') that could almost fly. Students would charge around on big open fields under the eye of an instructor until they were deemed to have enough of the feel of flying to move on to something that could actually leave the ground. While the scenes at training fields must have resembled stock-car demolition derbies, the system did have its virtues, not least of which was the minimizing of instructor casualties. It was largely to power these strange craft that Anzanis had been built in large numbers during the war.

But by 1919 the whole Anzani line was out of date as aircraft engines. Because the Anzani employed automatic intake valves, their power fell off rapidly with altitude; fuel consumption, at 0.66 lb/(hp)(hr) [0.3 kg/(hp)(hr)], was excessive, and they were, by post-war standards, crudely made. Examining an Anzani today makes one think of a power-mower engine rather than of an aircraft powerplant. By 1922 the entire Anzani line was modernized to use cam-operated intake valves, but even redesigned the Anzanis were not sophisticated engines. Although Anzani, in its fan-type engines, had been a pioneer in the use of link rods, its radials used a special type of construction that limited them to three cylinders per row. The Anzani main rod terminated in a cylindrical sleeve on the outside of which the other rods rode (Fig. 8-1). This forced it to go to 6 cylinders in two rows when more power than could be had from 3 cylinders was wanted.

The design of the cylinders was not susceptible to enlargement. They were of cast iron with integral heads and had the valves in line, with the exhaust in front (Fig. 8-2). Besides not allowing much air to get between the ports, this arrangement assures that the air reaching the intake port has been heated by passing over the exhaust port. It is better to keep the intake passages cool and, if necessary, cool the exhaust with air that has passed over the intake. The new line did have aluminum pistons and an oil pump, but the concept and finish remained generally outdated.

Not too much use was made of these engines. Farman built a small two-seater called at different times the Sport and the David (for comparison with Farman's big transport, which it labeled the Goliath) which was listed with the old automatic-intake-valve 45-hp six in 1923 and with the new 70-hp six in 1924, but *Jane's All the World's Aircraft* lists no other production aircraft with Anzani power. A few war-surplus 35-hp threes came to the United States, where they found their way into a number of rather unsanitary homebuilts and into one good one, Les Long's 1928 "Longster." Clyde Cessna used an Anzani six in the prototype of his model A, but production versions had Warners, Siemenses, and just about anything else available except Anzanis—which may tell the whole story.

In 1928 the entire line was again modernized to use slipper rod bearings, allowing up to 5 cylinders per row. An American named Brownback imported them and may have built some under license in Pennsylvania, but few were sold. Still

later, the Potez firm tried to revive the line, but the only plane to use the Potez derivative of the Anzani was Potez's own very attractive type 43 monoplane, with the new 100-hp six. Potez persevered to the extent of making a 60-hp three that seems to have been half of the six, and eventually most traces of Anzani ancestry disappeared. For example, the Potez 9Ba of 1933 obtained a surprising 250 hp from only 485 cu in. (7.95 L) at a moderate 2400 rpm and was apparently pure Potez. Unfortunately Potez's efforts went largely unrewarded, since by 1933 a growing Continental preference for inverted air-cooled inlines had all but eliminated the European market for small radials.

The Salmsons were much more popular than the Anzanis. The first of these had been introduced as early as 1919, and within several years a range of engines was available going from 12 hp to 500. Salmson was prone to confusing model-number changes, and so the best way to approach the firm's line is by describing its configuration and sizes with little reference to type designations.

One of the most visible and characteristic Salmson design features was its use of hairpin valve springs instead of the more usual coil type. Aside from this, the engines resembled the early Armstrong-Siddeley and Lawrance radials, having the same narrow angle between the valves. The valve gear was exposed, as was usual at the time. Steel cylinders with shrunk-on heads and conventional articulated connecting rods were used. By 1927 the Salmson line had shaken down to two cylinder sizes with 5-, 7-, and 9-cylinder engines available for each size. The smaller, called by then the AD series, had cylinders of only 20-cu-in. (330 cc) displacement. The tiny 9-cylinder AD9, called in the United States the "watch-charm engine," had a diameter of only 25.9 in. (658 mm) and gave 45 hp. Some AD9s were imported to the United States, where they replaced Szekeleys in American Eaglets and were a factory option in Aeromarine's license-built low-wing Klemm. As might be expected, these little engines were remarkably smooth (Fig. 8-3). A flight in a Salmson Klemm was about as pleasant an experience as any floater-type aircraft ever provided.

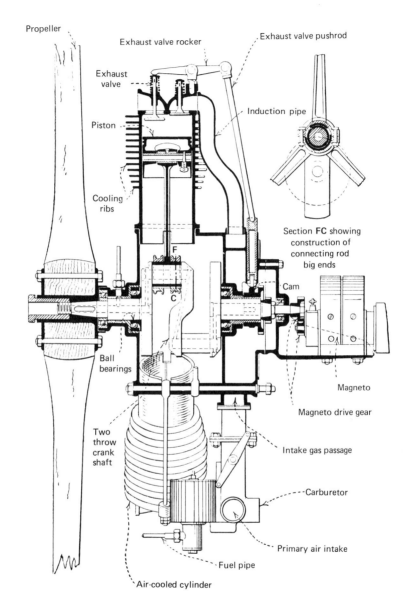

Propeller

Exhaust valve rocker

Exhaust valve pushrod

Exhaust valve

Piston

Induction pipe

Cooling ribs

Section **FC** showing construction of connecting rod big ends

Cam

Ball bearings

Magneto

Magneto drive gear

Two throw crank shaft

Intake gas passage

Carburetor

Primary air intake

Fuel pipe

Air-cooled cylinder

FIG. 8-1 Construction of Anzani 6-cylinder, two-row radials prior to 1922. The automatic intake valve and the peculiar design of the connecting rod bearings are clearly shown. Note the jog, or offset, built into the crankshaft and just clearing the thin connecting rods, to minimize the offset of the rear cylinders. *(Courtesy of NASM-Smithsonian.)*

FIG. 8-2 Cylinders of a post-1922 Anzani 6-cylinder radial. Compare with Fig. 8-1; there is a strong feeling of "afterthought engineering." The rocker arms and pushrods for the exhaust valves have, it seems obvious, been grafted onto the old design with the minimum possible amount of change. *(Courtesy of the author, taken at NASM-Silver Hill.)*

The larger AC series came as a five of 65 hp, a seven of 95, and a nine of 129. They powered a wide range of French lightplanes, almost all two-seaters. Europe built none of the three-seat biplanes so common in America. These had been developed in accordance with the characteristics of the ubiquitous OX-5 and with the need, in the absence of government subsidies for private flying, to support one's flying by taking up passengers for joyrides. Perhaps the most consistent Salmson user was Morane-Saulnier. This old firm built parasol monoplanes in many styles, almost all with swept-back wings, between 1915 and 1935; most civilian models used Salmson power.

There were only three other small French radials. Renault built a 120- to 130-hp seven from about 1930 to 1934, Lorraine a five of 120 hp during the same period, and Hispano-Suiza a five, basically the Wright J-6-5, of 150 hp. Few of these were sold. As far as the small radial was concerned, France was Salmson country.

Small British Radials

The British built a fair number of small radials despite their general preference for the air-cooled inline. British small radials displayed a strong tendency toward the use of small numbers of rather large cylinders. A notable example of this was the Bristol Lucifer, built during the mid-twenties. It was a 3-cylinder engine using destroked Jupiter cylinders and gave 120 hp, making it the most powerful 3-cylinder engine ever built. Its excessive vibration prevented it from becoming popular. Armstrong-Siddeley evolved its Mongoose from the Jaguar in the same fashion, but, as the Jaguar cylinders were smaller, the 125-hp Mongoose was a five. The Mongoose was fairly widely used, notably on the well-liked Avro trainers that succeeded the wartime 504, but the most popular Armstrong-Siddeley was the 65-hp Genet, also a five. Some Genets came to the United States (Fig. 8-4).[1]

FIG. 8-3(a) The Salmson AD9, 45 hp from 9 cylinders! The engine in the background is an old automatic-valve Anzani. *(Courtesy of the author, taken at Old Rhinebeck Aerodrome, New York.)*

[1]Under peculiar circumstances. A hundred engines had been ordered—optimistically—to power Driggs Dart 2 biplanes, but only 20 Dart 2s were built; so the other 80 came on the market.

FIG. 8-3(*b*) Hairpin valve springs on a medium-size Salmson radial. *(Courtesy of the author, taken at NASM-Silver Hill.)*

Not until the first Pobjoys came out in 1928 was there a small British radial designed as such from scratch. The Pobjoys were interesting engines. The geared models used spur-type reduction gears,

resulting in the propeller shaft being offset above the engine's centerline. They were also unusual in taking advantage of their small size and reduction gears to run at very high rpm.

The first Pobjoy was the "P-type." It gave 65 hp from 151 cu in. (2.47 L) at 3000 rpm geared down to an efficient 1570 at the propeller. It had a diameter of only 25 in. (635 mm). The "P-type" used a peculiar ignition system. Ordinarily a 7-cylinder engine uses a seven-unit magneto running at half engine speed. Each unit of the "mag" takes care of its own cylinder. The first Pobjoys, however, used "single-cylinder" magnetos running at 3.5 times engine speed. The same unit fired all cylinders. This gave trouble; in addition, the exposed valve gear wanted watching. These faults were eliminated in the 1931 "R-type" series, which was, in addition, larger at 173 cu in. (2.84 L) and 75

FIG. 8-3(*c*) Morane-Saulnier sport-trainer, about 1929, with Salmson AC 9 engine. Note that propeller rotation, as in all European engines, is opposite to that of American practice. One of these Moranes, faked-up with German insignia, was crashed in the final scene of "The Blue Max." *(Courtesy of the author, taken at Old Rhinebeck Aerodrome, New York.)*

FIG. 8-4 Armstrong-Siddeley Genet Major, about 115 hp. The 5-cylinder version was more popular and was one of the few foreign radials used to a significant extent in the United States. *(Courtesy of Rolls-Royce Limited.)*

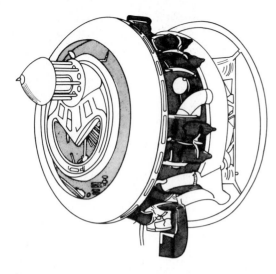

FIG. 8-5 Pobjoy Niagara radial, the only radial engine ever to use spur-type reduction gears. *(Courtesy of the author's collection.)*

hp. In 1934 the "R-type" became the Cataract, if geared, and the Cascade in the 70-hp direct-drive version. Also in 1934, the Niagara, the definitive Pobjoy, came out (Fig. 8-5). The Niagara was sold fully cowled, had enclosed and automatically lubricated valve gear, and delivered 90 hp at 3500 rpm.

Pobjoy claimed that its engines were used in many different countries and in a wide range of aircraft, but this was somewhat misleading. They tended to go into experimental and small-production planes rather than in widely used ones.[2] Recently an enthusiast who was restoring a Pobjoy engine was quoted as saying that no two Pobjoys seem to have been entirely alike, no doubt something of an exaggeration but nonetheless strongly suggestive of semi-custom manufacture.

The classic Pobjoy-engined plane was probably the Comper Swift single-seater, but Niagaras went into the strangest of all small transports. This was the Short Scion Senior,[3] a four-engine nine-passenger monoplane. In an interesting example of

cautious engineering, the design of the Short Sterling bomber was first checked out in a four-Niagara human-carrying model based on the Scion Senior.

Apparently there was only one other small British radial line. Wolseley Motors, part of the Nuffield empire (MG, Morris, etc.) and a builder of Hissos under the style of Wolseley Viper during the First World War, came back to the aircraft field in 1933 with two radials. These were conventional engines, perhaps a bit outdated with their narrow angle between the valves. They used internal spur reduction gears (Fig. 8-6) for 2 years and then switched to the epicyclic type. Wolseley radials were unusual in using Monel Metal valve seats. Like so many others, it built a seven and a nine using the same cylinders, the 145-hp A.R. 7 and the 180- to 203-hp A.R. 9. By 1935 the Wolseleys had zodiacal names like "Aries," "Aquarius," and so on, but the British market did not respond, and the firm left the business in 1937.

Czechoslovakia, Germany, and Belgium

The Czechoslovakian firm of Walter was the pioneer of small radials in Central Europe. Its first engines were built in 1922 and combined some advanced features with others that were decidedly retrograde. Walter used steel cylinders which it dipped in molten zinc to provide bonding for cast-on aluminum cooling fins but negated this advance by employing cast-iron heads. Valves were set

FIG. 8-6 Internal spur reduction gears. Tooth stresses are low because several teeth are sharing the load, but deflection is a problem because both gears are "overhung" rather than "straddle-mounted." Propeller shaft is offset slightly from the crankshaft but nowhere near as much as with ordinary spur gears.

---

[2]Or, as in the case of the early Bellanca 14, as a not-too-popular option.

[3]I'm not making this up. As shown by the Jowett Javelin Jupiter and the Humber Super Snipe, the English see absolutely nothing silly about such names.

vertically. The fins on the heads were annular and looked like a continuation of those on the cylinders, giving the engine a neater look than one with the better-cooled splayed valves. Walter built threes, fives, sevens, and nines with ratings from 55 hp to 135 hp, all with the same cylinder design.

In 1928 all these were replaced by newer designs and supplemented by a larger series. The small series comprised the 3-cylinder, 60-hp Polaris, the 5-cylinder, 90-hp Vega, the 7-cylinder, 115-hp Venus, and the 9-cylinder, 150-hp Mars. The larger line was really out of the small-engine category at 260 to 345 hp for the 7-cylinder Castor and 365 to 420 hp for the 9-cylinder Pollux, but it also included a 5-cylinder companion, the 185/210-hp Regulus. The second series Walters were of more conventional design than the earlier models, but the angle between the valves was still too narrow for best cooling. The Castor and the Pollux were popular engines, but of the smaller Walters, only the Mars saw extensive use. By the late twenties the air-cooled inline had become the engine of choice, in Europe, for applications requiring 125 hp or less, and the smaller radials went into eclipse (Fig. 8-7).

The Siemens engines were, in many ways, the German counterpart of the Walters. Siemens engines were similar in size and appearance to the Walters, leading some writers to assume a connection between the two, but their construction and dimensions differed. Siemens used aluminum cylinders and heads with steel liners and put the valves in a fore-and-aft line with the exhaust to the rear. As mentioned in connection with the Anzani line, this is a poor way to cool the exhaust valve. It can be done, but it imposes a low limit on the amount of power that can be safely had from a given displacement. Each cylinder had a pair of coaxially mounted rocker arms, operated by pushrods placed one behind the other [Fig. 8-8(a)].

This heroic design vanished after 2 years of production. New and better cylinder designs of the same 3 15/16-in. (100-mm) bore and 4 23/32-in. (120-mm) stroke were introduced in 1925. Since the same crankcase was used, some ingenuity was called for in designing the new valve gear, and Siemens's solution was to use Z-type, or "long-bearing," rocker arms [Fig. 8-8(b)]. The new heads

FIG. 8-7 A late small Walter radial, the 240-hp Bora of the mid-thirties. *(Courtesy of the National Archives.)*

cooled much better, and output rose from the original 11 hp per cylinder to 14. Then, between 1928 and 1934, all three engines were given enlarged bores [to 4 ⅛ in. (105 mm)] and conventional fore-and-aft rocker arms. The same basic lower end was retained and the pushrods were splayed out at a wide angle to engage the exposed rocker arms. Horsepower rose again, to about 16½ hp per cylinder, for the five and the seven, but the nine was dropped.

The Siemens radials were the preeminent lightplane and military trainer engines in Germany until the air-cooled inlines took over first place. A few were imported to the United States, where they were available on Cessnas and Wacos, but only Teutonophiles bought them. If nothing else, the spare parts situation would make it hard for them to compete with domestic engines in the same horsepower class.

While the line was given some improvements from time to time, progress was not rapid. Even by 1936, when the name was changed to Bramo (see Chap. 7), only the seven, by then enlarged to 160

FIG. 8-8(a)   Early type Siemens head and valve arrangement. (*Courtesy of NASM-Smithsonian.*)

hp, had achieved the luxury of covered valve gear [Fig. 8-8(c)]. The seven became moderately well known here as the powerplant of the incomparable aerobatic biplane, the Bucker Jungmeister. Originally the Jungmeister had used the Hirth HM-6 inverted inline, and a comparison between the Bramo and the Hirth is revealing. The Hirth gave 135/140 hp at 2400 rpm and weighed 330 lb (150

kg); the Bramo gave 160 hp at 2200 for 298 lb (135 kg). With its smaller frontal area, the Hirth might provide more speed in a given aircraft, but the Bramo's lower weight, higher power, and slightly lower rpm would result in distinctly better climb. In addition, an aerobatic aircraft like the Jungmeister would benefit from the lower polar moment of inertia conferred by the radial's shortness. In retrospect, there is little doubt that the pattern of using inlines for family and business planes and using radials for trainers made considerable sense.

Two other Continental small-engine builders used the same system of valve actuation as the early Siemenses. One was Renard, in Belgium, who made fives, sevens, nines, and an eighteen between 1927 and 1935. Renards had exposed pushrods and rocker arms like the Siemenses, but Skoda, in Czechoslovakia, the other user of this system, enclosed them. A neat and attractive engine resulted, but, in competition with the better Walters, sales were few in the limited Czech market, and the line—by then called "Avia"—came to an end by 1935.

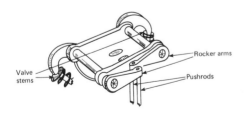

FIG. 8-8(b)   Z-type, or "long-bearing," rocker arms as used by Siemens.

FIG. 8-8(c)   Bramo seven, the Jungmeister's engine. (*Courtesy of NASM-Smithsonian.*)

Italy and Spain

The first Italian maker of small radials was Fiat, which began with water cooling (see Chap. 6) but which went to air cooling in the mid-twenties. There were three small Fiats, all sevens: the A 50 of 90/106 hp, the A 53, in 1930, of 120 hp, and the 1933 A 54 of 138/150 hp. The small radials could never have been anything but a sideline for Fiat. This major firm seems to have tried every possible type of aero engine—big liquid-cooled engines, small air-cooled inlines, big radials under foreign license, and even an aircraft Diesel—and quickly dropped those that failed to pay their way (Fig. 8-9).

Stabilimenti Farina, operated by the uncle of the well-known Pinin Farina and itself better known as an automobile body builder, built a 7-cylinder radial of 68 to 100 hp between 1929 and 1935. It was called Algol in direct-drive form and Alioth if

FIG. 8-9   Fiat A 50, 7 cylinders, about 1928. Although the rocker arms and pushrods are enclosed, they are not automatically lubricated from the engine's central lubrication system; the Tecalemite grease fittings can be clearly seen. This was one of the very few radials to use clips, instead of studs, to retain the rocker arm covers. *(Courtesy of Centro Storico Fiat.)*

geared. Its most interesting feature was a combination of roller and slipper bearings for the connecting rods, which must have made for a very smooth engine. In 1936 the firm added a big five, the T. 58 of 135/142 hp. While there were no civilian sales, the government bought a fair number for training planes, many of which were furnished to the subsidized flying clubs. Farina production of aircraft engines ended shortly before the war; I believe Sr. Farina was killed in an automobile race.

Alfa-Romeo, like Fiat, was better known for large engines, building Bristols under government license (see Chap. 7) and the Armstrong-Siddeley Lynx under its own arrangement, but it did make one small radial of its own design. Called simply the Alfa-Romeo D, it was a conventional nine with a narrow angle between the valves and gave 225/245 hp (5 more when supercharged). In the 1935 D-2 version it was rated at 290/310 hp when naturally aspirated and, oddly, at only 275/310 in the supercharged version. As this power range overlapped that of its license-built Lynx, it is hard at this distance in time to see the point in the whole exercise. These engines were little used, even Alfa's Romeo-Meridionali subsidiary preferring the Lynx for its lighter planes.

In Spain the Elizalde firm made a valiant effort. Its Dragon series were biggish engines, but the 5-cylinder Dragon V of 172 hp falls into the small-engine category. This firm had been a builder of trucks and buses, and there is a cloudy tale to the effect that it once built the world's largest-ever car. It brought considerable professionalism to the job of building Spain's first homegrown aircraft engine. Unfortunately, the Elizalde, like so many promising things in republican Spain, fell a victim to the Italo-German Fascist invasion that we like to miscall a civil war.

The Smaller Radial in America

Until the supply finally ran out in 1928, the OX-5 stood in the way of American small-engine development. Most of our private-owner aircraft of the twenties were variations on the same theme: a three-seat biplane powered, if that is the word, by an OX-5, with a 150-hp Wright-Hisso available as an option for those who could afford a real engine. These were not minimum airplanes, gross weight varying from about 1750 lb (795 kg) for the Bird to perhaps 2500 lb (1136 kg) for the Alexander Eaglerock. This implied that designers could go in either of two directions when the time came to develop OX-5 replacements to go into these planes. One would be to make a lighter engine of power about equal to that of the OX-5; the other would be to make a more powerful engine of about the same weight.

Either approach involved problems. When a 275-lb (125-kg) Kinner replaced the 400-lb (182-kg) OX-5 in the American Eagle, stability changed from positive to, at best, neutral. In addition, the extended nose used to move the lighter engine forward added side area ahead of the center of gravity, to the detriment of directional stability. On the other hand, if a larger engine was installed, there was some danger that the added performance would overstress the airframe. Fortunately, most airframes of the time were amply strong and so aerodynamically dirty that this seldom happened.

In fact, often there was surprisingly little gain in speed when a more powerful engine replaced an OX-5. There were two reasons for this. One was that most OX-5 installations were fairly cleanly cowled, and drag was less than when an exposed-cylinder radial was used. The other was the fact, discussed in Chap. 4, that the OX-5's low speed of 1400 rpm resulted in good thrust from wooden propellers. Of course, climb improved, which was easy because no OX-5-powered plane, not even the light and efficient Bird, could ever climb for sour apples.

Unless one counts the Lawrance J-1, which, at 180/200 hp, was on the borderline of the small-radial category, we had only two small radials before 1927. One was Lawrance's 3-cylinder model L of 60 hp, built about 1919. This was large enough for a light two-seater had there been any market for such craft. It was used on the Sperry Messenger military courier plane and on Giuseppe Bellanca's single-seater biplane, but it couldn't compete in the market with surplus OX-5s at $290. Its importance lay in its role as the ancestor of the Whirlwind.

The other American radial was a decided oddity. A Texas company called Tips and Smith converted the surplus 80-hp American-built Le-Rhone rotary into a 120-hp static radial. It was offered at $500 FOB Houston, and a few were used in various conversions, but it could not have been a good engine. It employed the original LeRhone cylinders and heads, and since these had had to spin at 1200 rpm to get rid of the heat generated in producing 80 hp, there was no possible way they could have kept cool while standing still and producing 120 hp.

First of the larger OX-5 replacements was the Continental A-70 (Fig. 8-10) of late 1928. Continental was a Detroit-based maker of automobile and truck engines who supplied many of the nation's smaller car builders. By the late twenties this "assembled car" market was shrinking, and so it was natural for Continental to move into the growing aircraft field. Interestingly enough, before building the A-70, Continental had built and

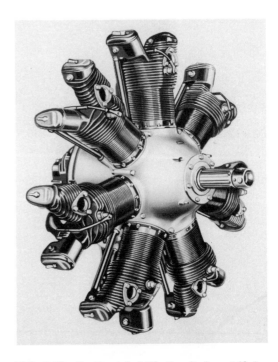

FIG. 8-10   Continental A-70, 7 cylinders, 160 hp. *(Courtesy of Teledyne Continental Motors.)*

shown—but not flown—a Burt-McCollum single-sleeve-valve radial.

The A-70 was a conventional design—unless one considers the placing of the cam ring and pushrods at the rear a significant departure—and was smooth and reliable from the beginning. It sacrificed maximum power in the interests of simplicity by using an exhaust-heated muff to warm the incoming mixture for good distribution, in preference to a mixing blower. In its original form it gave 170 hp at 2000 rpm. It was used in some Wacos—my own first flight, at the age of 12, was in a Continental-powered Waco—but otherwise it took a few years for sales to gather momentum. Its best-known application was in certain models of the classic "Stearman" trainer of World War II; I believe the Naval Aircraft Factory's N3N version used it exclusively.

Continental's difficulty was that many of its competitors in the engine business had ties with airframe manufacturers. Curtiss, not yet merged with Wright in 1928, is a case in point. Curtiss was to some extent pushed into the small-radial business. Its three-seat Robin monoplane had been designed around the OX-5, but it was selling so well that Curtiss found it worthwhile to design a new engine for it when the OX-5 well ran dry. The new engine was the Curtiss Challenger (Fig. 8-11), a two-row six of 180 hp. It weighed 420 lb (191 kg), almost exactly the same as an OX-5 with water and radiator, and, with a specific consumption of 0.5 lb (0.227 kg), it was gratifyingly economical. There was only one trouble with the Challenger: it was one of the worst-vibrating engines ever built. It is the only engine I know of that was frequently replaced for no other reason than that owners refused to put up with its roughness. Douglas Corrigan's "wrong way" transatlantic Robin was typical in that a previous owner had replaced the Challenger with a J-6-7.

Probably the two-row construction contributed to the vibration problem, since a two-row six is actually two threes, but more of the trouble probably came from the "oversquare" cylinder dimensions of a 5 ⅛-in. (130-mm) bore by a 4 ⅞-in. (124-mm) stroke. A relatively large and slow-turning engine ordinarily gains in smoothness by the use of a long-stroke design and vice versa. If

FIG. 8-11 Curtiss Challenger, America's only 6-cylinder radial. *(Courtesy of NASM-Smithsonian, from the Sherman Fairchild Collection.)*

these unusual dimensions were adopted in an effort to hold down the diameter, the effort was misdirected, because the Challenger, at 42 ⅝-in. (1080-mm) diameter, was at least as large as most engines of similar power.

While a few Command-Aires used the Challenger, only Curtiss itself employed it in any quantity. Besides being used in the Robin, which became a four-seater when Challenger-powered, it was used in Curtiss's Fledgling. This was a trainer for the Navy and for Curtiss's own flying schools of the late twenties. It was large and clumsy-looking but strong and safe. The Fledgling was the last two-bay biplane built in the United States. Two survive, one with its original Challenger; this has won a Grand Award at the Antique Aircraft Association's Iowa Fly-In. Between these two aircraft, there was enough production to call for the building of a very respectable number of Challengers. Every aircraft museum has one because nobody could stand to use one long enough to wear it out.

Lycoming was another engine builder that gained through its corporate connections. It was part of the empire put together by the remarkable Erret Lobban Cord, of Auburn-Cord-Duesenberg car fame, and as such was the preferred engine supplier for Stinson. Since Stinson was in turn a source of equipment for Cord's airlines, the Lycoming radial went into Stinson Trimotors[4] as well as into the popular Detroiters, Reliants, and Juniors. One of these, the low-wing model A trimotor of 1934, had an unusual problem with its center engine. While the Lycoming was not an especially dirty engine, the forward-sloping windscreen on the A so effectively caught whatever oil did leak out from under the cowling that pilots on American Airlines' secondary routes, where the A was mainly used, became adept at blind landings long before radar was invented.

The Lycoming, like the Continental, was conventional and sound in design. It had 9 cylinders and hence was smoother than such rivals as the Continental, Wright J-6-7, and Jacobs, and was fairly economical of fuel. It originally gave 210 hp, went as high as 260 in the R-680-5 version, but was most popular in the 220-hp model. It was the basic engine for the "Stearman" trainer, the designation of PT-13 applying only to Lycoming-powered "Stearmans." From all accounts it rivaled the Wasp Junior in being as reliable as a hammer. It just never quit.

Wright, though not an airframe builder until it merged with Curtiss, had a considerable competitive advantage as a result of the sterling reliability demonstrated by the original Whirlwind's many long-distance records. The 9-cylinder J-6 Whirlwind, at 975 cu in. (15.98 L) and 300 hp, was well out of the small-engine category, but Wright also built 5- and 7-cylinder versions. The five was called at first the Whirlwind 165, for its horsepower, and later the J-6-5 and, for its displacement, the R-540. At a weight of 380 lb (172 kg), it would have given good performance, but vibration may have been a drawback in competition with the 7-cylinder engines available in the same power range; it was not a popular engine. The 7-cylinder J-6-7, giving 240 hp and weighing only 30 lb (14 kg) more than the Challenger, was much the better seller of the two. Many well-known

aircraft of the period of around 1930 used it: various Wacos, the four-place Ryan Brougham (the six-place Ryan used the J-6-9), the Stearman C-3, and some of the Travelairs. One took Douglas Corrigan to Ireland.

The fourth major American radial in this class was the Jacobs. Jacobs began in Camden, New Jersey, in 1930 with a 55-hp F-head three and a 150-hp seven. The 3-cylinder engine had a short production life, but the seven founded a long-lived line. Its output rose to 170 hp in 1931 and, slightly enlarged as the LA-2 model, to 195 in 1932. The LA series Jacobs used exposed Z-type rocker arms; they were the only U.S. engines to employ this dubious design feature. This was eliminated in the 1934 L-4, which used enclosed fore-and-aft rocker arms of classic design. The L-4 developed 225 hp from 757 cu in. (12.4 L). The 285-hp L-5 was added 2 years later. At 831 cu in. (13.6 L) the L-5 was the largest American seven until the post-war Cyclone Seven was built, and it is easy to understand why its World War II nickname was the "Shakin' Jake."

In the market as in design, the Jacobs was a late bloomer. While used to some extent during the late thirties—the Beech Staggerwing and the very advanced Spartan Executive were Jacobs-powered—it did not come into its own until the Second World War. Cessna used two in the "Bamboo Bomber" UC-71 and evidently became a convert to Jacobs power, since those exemplary airplanes, the Cessna 190 and 195, were Jacobs-powered.[5] One peculiarity showed up in the Cessna installations. The Jakes used coil ignition instead of magneto, and the distributor caps were prone to cracking. When this happens, the spark travels to the wrong cylinders and the engine runs rough. The cure was to have the cap undergo a bakeout in a vacuum oven and then impregnate the crack with insulating varnish, after which it would run forever without giving further trouble. The only difficulty was that this was illegal. What the government wanted you to do was replace the cracked cap with a new crack-prone cap (Fig. 8-12).

---

[4]The Stinson Trimotors were the preeminent small American multiengine transports until the coming of the little Lockheed 10s and 12s.

[5]True, you couldn't see over the engine, but nothing's perfect.

FIG. 8-12 Jacobs R-755 (L-4 MB). A well-developed workhorse of an engine. (Courtesy of the author, taken at Glenn Curtiss Museum, Hammondsport, New York.)

Engines at the lower end of the OX-5 replacement range began to appear, in a tentative way, as early as 1927. In that year, "Eddie" Rickenbacker and former Wright Field Chief Engineer Glenn D. Angle teamed up to make the Detroit Aircat, of 5 cylinders and 60 hp, and the first of the Kinners came out in Glendale, California.

The history of the Aircat and its descendants is complex. The original engine was used in the once-well-known Monocoupe, built in Moline, Illinois, by Clayton Folkerts and Don Luscombe. Dissatisfied with the Aircat's unreliability and with the factory's difficulties in making deliveries on time, Mono Aircraft got the Velie automobile firm, across the river, to build an aircraft engine for it. The 60-hp Velie incorporated such needed improvements as an aluminum head instead of the Aircat's iron one but otherwise copied the Aircat so closely that Velie had to pay damages in a patent suit. Then two senior members of the Velie family died during the same year, and the firm was folded into the John Deere tractor business, with which it had some connection. The engine was turned over to a Mono Aircraft subsidiary and became the

Lambert. In its final form the Lambert [Fig. 8-13(a)] was a good-looking, chunky little engine with prominent cylinder hold-down studs and pushrods at the rear. It was made in two sizes, 55/60 hp and 85/90 hp. As nobody but Mono ever used the Lambert, when the Monocoupe finally became obsolete after 10 years of production, the Lambert died with it.

In the meantime the Aircat had evolved into the LeBlond [Fig. 8-13(b)]. The LeBlond firm was a famous Cincinnati machine-tool builder; older machinists remember the LeBlond lathe as one of the best. LeBlond bought the services of Glenn D. Angle and the rights to the Aircat. It was redesigned to have enclosed valve gear, and additional models were added. The enclosed valve gear was not an unmixed blessing, because throughout their production life the LeBlonds and their successors had a tendency to swallow valves unless carefully maintained. Exposed valve gears might have encouraged the paying of closer attention.

The expanded line under the LeBlond name included iron-head fives of 60 and 70 hp, an iron-head 90-hp seven, an aluminum-head five of 85 hp, and an aluminum-head seven of 110 hp. As a business proposition, however, the airplane engine game did not live up to the hopes of Richard LeBlond, the company's president. Selling in the low end of the horsepower range, the LeBlonds met stiff competition from Szekeley's and Aeromarine's low-priced 3-cylinder radials. The opposed engines were not yet competitive as long as they produced less than 40 hp, but it is unlikely to be coincidence that Continental's announcement of the A-65 and LeBlond's sale to Rearwin came within a year of each other.

Rearwin had been LeBlond's main customer, and some of its planes had been designed around LeBlond engines. It changed the name to Ken-Royce, for Mr. Rearwin's two sons, and built them for 2 or 3 years until the outbreak of the Second World War. Then the firm sold out to Commonwealth, which dropped the engines and the radial-engine aircraft to concentrate on those powered by opposed engines. Interestingly, the engines could apparently still be labeled "LeBlond" if a customer wished. Bellanca had used the LeBlond in the plane that eventually became the Cruisemaster, and it continued to use "LeBlond" in its advertising

FIG. 8-13(a) Velie M-5, essentially the same as the Lambert 55/60-hp model of 1928, on a museum display stand. The power plant of the classic early Monocoupes. *(Courtesy of the author, taken at Glenn Curtiss Museum, Hammondsport, New York.)*

even after the engine became the "Ken-Royce" elsewhere.

A better engine at the low end of the OX-5 replacement market was the Warner Scarab. This was a 7-cylinder engine of conventional design and with an introductory output of 110 hp. It received a sensational introduction at the 1928 National Air Races, Cessnas and Monocoupes with Warner power dominating their classes. In 1929 it got a strengthened crankshaft, and output went to 128 hp. A 90-hp five was added in 1930, but it was never as popular as the seven. In 1932 the Super Scarab of 145 hp was added and, in 1938, a 165-hp version. These engines were much smoother than such fives as the Kinners and found their way into a different class of aircraft. Warners powered the beautiful Cessna Airmasters of the thirties and the Fairchild 24, all comfortable four-place cabin planes. Except for a few that were used in Fairchilds, Warners did not fit into the war production program; production was phased out during the war and was never resumed.

The first Kinner was the K-5 of 372 cu in. (6.1 L), a 100-hp engine of somewhat odd appearance. It had 5 long, slender cylinders on a small-diameter crankcase; the pushrods were behind the cylinders, making it look like it was all cylinder and not much else. This appearance and a heavy, throbbing vibration remained Kinner characteristics throughout the long life of the make. Kinners were the last engines in the world that any sane person would use in an airplane with any pretensions to sleekness or elegance, but they were great engines just the same—rugged, durable, reliable, and economical (Fig. 8-14).

Power went to 104 hp by 1930, then to 110, then back to 100 for the rest of the K-5's production life. Some airplanes are listed as having a 90-hp Kinner, but as far as the manufacturers were concerned there was no such thing. The line grew to include the 125-hp B-5 in 1930, the 210-hp C-5 in 1931, and the 165-hp R-5 in 1934. The K-5, B-

FIG. 8-13(b) LeBlond radial—the final development of the series that began with Aircat. *(Courtesy of NASM-Smithsonian.)*

FIG. 8-14(a) A big (160-hp) Kinner R-540 (R 55) cutaway to show the works. Note that the spiky appearance of the Kinner is due to the fact that the entire cylinder is externally mounted on a crankcase protrusion or boss, not because the cylinders themselves are exceptionally long. It was more common design practice to use a larger-diameter crankcase with the cylinders protruding slightly into it (spigoted in). (Courtesy of the author, taken at Glenn Curtiss Museum, Hammondsport, New York.)

on each separate cylinder might have been effective.

At the very bottom of the horsepower scale were the 3-cylinder engines. Whereas European threes were usually small versions of existing fives and sevens, the American threes were products of firms specializing in the type. The abortive Lawrance aside, there were two such engines that enjoyed some degree of commercial success.

First in point of time were the Szekeleys, beginning in 1928 with the 40-hp SR-3. This was an iron-cylinder engine that used slipper rods and delivered its power at a useful 1800 rpm. In 1929 an L-head version was added; this retrograde step reduced output to 30 hp. Americans were not interested in such low-powered engines before the Crash, but the timing was right for the 45-hp type O when it came out in 1931. The Depression was forcing many fliers to seriously consider airplanes that they would have scorned as "power gliders" a few years previously. Curtiss-Wright's problems exemplify the change that was going on. This major

5, and R-5 were widely used, but not the C-5 or Kinner's only seven, the C-7. From beginning to end, Kinner meant a 5-cylinder engine to everyone. Kinners of from 100 to 160 hp went into Fleet biplanes, into Birds, and into Kinner's own strut-braced monoplanes as well as into various less well known craft (Fig. 8-15). During the war, Ryan had to commit the aesthetic crime of substituting a Kinner for the Menasco in the S-T. As might have been expected, speed dropped and climb improved. Had the Kinner been cowled, speed might have been improved, but as this was a military trainer, under style of PT-22, accessibility for maintenance purposes had to come first. In fact, I know of no cowled Kinner installations. Either a Townend ring or a NACA cowl would have looked strange on an engine with such wide spaces between its cylinders; perhaps a Russian-style "helmet" cowl

FIG. 8-14(b) Kinner 100-hp K-5 in Gee Bee (!) biplane. Except for the feature of side-by-side seating, rare for an open-cockpit biplane, the Gee Bee typified the sort of plane for which Kinners made an ideal powerplant. (Courtesy of the author, taken at Bradley Air Museum, Windsor Locks, Connecticut.)

FIG. 8-15 Brunner-Winkle (later made by Perth Amboy Title Company, surely the most oddly named aircraft maker this side of China) Bird, 100-hp Kinner engine. Although not as able to take abuse as a Waco, this was the most efficient of the American three-place biplanes. *(Courtesy of NASM-Smithsonian.)*

firm had an extensive line of civil aircraft which had sold well as late as 1930 but which were dying on the market by 1932. Out of sheer desperation, it bought the rights to a little amateur-built single-seat pusher and enlarged it to a Szekeley-powered two-seater. This little plane, called the Curtiss-Wright Junior, was credited with saving the great Curtiss-Wright company. I can say that, to a kid whose previous flying had involved sitting surrounded by wires and wings in a vibrating Waco, the Junior was pure joy. The view was magnificent, and most of the noise was left behind. It was slow, though, and probably didn't look like a real airplane to most people; it disappeared quickly when planes of the Cub type became available.

Buhl was another Szekeley user but was less fortunate than Curtiss-Wright. Its Bull Pup, an Etienne Dormoy design, combined a very modern monocoque aluminum fuselage with old-fashioned wire bracing for the wood-structure wing. Unfortunately, this attractive midwing monoplane was a single-seater, always a good way to make sure that sales are limited, and the Buhl firm's career as a plane builder ended with the Bull Pup. As a reminder of the importance of low rpm to good performance, it is worth noting that today a restored Bull Pup is getting about the same performance from a 2350-rpm Continental A-65 as it originally got from the Szekeley's 45 hp at 1750. It's the prop that hauls the airplane along and cubic inches that swing the prop, and the Szekeley's 190 cu in. (3.11 L) could swing more prop than the A-65's 171 cu in. (2.8 L) can.

Another three of a few years later was the Aeromarine. This was a smaller but more finished-looking engine than the Szekeley. It gave 55 hp for takeoff at 2400 rpm and 50 at 2150 for cruising, from 160 cu in. (2.62 L). It was well enough designed, having, for instance, aluminum heads from the start, but its use of articulated connecting rods made it a rougher engine than the Szekeley. Both Szekeleys and Aeromarines were used on a number of conventional high-wing monoplanes of the same type, but of lesser fame, than the Cub. I remember an American Eaglet that demonstrated the severity of the Aeromarine's vibration when installed in a light airframe. It had a spring-vane airspeed indicator mounted on the wing struts, but the vibration was so severe that the struts themselves became a blur and the airspeed indicator could hardly be seen, much less read. Oddly enough, Aeromarine never advertised its own engine as an available powerplant for the Klemms it was building under license, but in fact some Aeromarine Klemms did employ this engine. The Klemm's thick cantilever wing and plywood structure damped the vibration far better than did the

FIG. 8-16 Aeromarine 55-hp radial of the early thirties, a reliable but somewhat rough little engine. Note the coarsely pitched head fins—perfectly practical on small cylinders because of the relationship between surface and volume mentioned earlier. (*Courtesy of NASM-Smithsonian.*)

composite construction of the Eaglet or the Porterfield, another occasional user (Fig. 8-16).

Aeromarine did not make this engine for long, but it was revived in 1938 as the Lenape Papoose. By then the opposed engines had grown beyond their earlier 40-hp maximum and were well on their way to market domination, and the Papoose had no chance whatever.

There were a few others, but none was produced in appreciable volume. On the West Coast the Axelson firm, builders of big lathes, took over an engine called the Floco in a partly developed state and brought it to market readiness. A 150-hp seven, it was used on a few Colorado-built Eaglerocks around 1930. W. B. Kinner left Kinner Motors in 1935, taking with him the rights to the Airster plane, and shortly reinvented the Kinner five as the Security. In Pennsylvania the Anzani was built (or merely assembled?), at first as the Brownback and later as the Light. The 7-cylinder Comet came from Wisconsin in 1930. Glenn D. Angle tried again, late in the period, with an engine carrying his own name, but the war intervened. Charles B. Lawrance built a tiny 25-hp five that ended up powering auxiliary generators on the B-29.

All these were conventional engines, but there were at least two that were a little different. One was the McClatchie Panther, a Compton, California, product. This was the most promising L-head radial ever built (not much of an endorsement, perhaps). It had 7 cylinders and gave 150 hp from the remarkably small diameter of 36 in. (914 mm). While the desire to minimize bulk was undoubtedly the chief reason for the adoption of the L-head configuration, McClatchie's advertising made much of the then-true fact that broken rocker arms were a principal cause of engine trouble. Most L-head engines are less efficient than comparable overhead-valve types, but the moderate specific consumption of 0.56 lb (0.254 kg) was claimed for the Panther. It never sold, and the reason why is impossible now to tell (Fig. 8-17).

Another exploration into the mildly unconventional was the Kimball Beetle, a few of which were made in Naugatuck, Connecticut, in 1931. The Beetle was a 5-cylinder F-head engine of 135

FIG. 8-17 McClatchie Panther, 150 hp. A serious attempt at making an efficient L-head radial engine. Notice the six radial fins that break the smooth line of the head fins over each valve. These are removable plugs. The head was made in one piece with the valve ports and these plugs had to be removed to get at the valves; an old design technique, going back to the days when water-cooled L-head engines had their heads cast in one piece with the cylinders. *(Courtesy of NASM-Smithsonian.)*

hp. The F-head can give a useful compromise between the turbulence imparted to the fuel charge in an L-head and the better breathing of an overhead-valve type. The Kimball claimed good fuel economy, and its weight, despite the use of cast-iron heads, was moderate for its power, but its diameter was excessive. At 45 in. (1140 mm) it was as bulky as some engines of several times its power. It was listed as an optional engine for a few planes, but the combination of the Depression and inadequate financing did it in before it got a chance to prove itself (Fig. 8-18).

Dim Glimmerings from the East

The small radial has no post-war history in the West. The opposed engine achieved rapid domination after the war. The only radial-engined planes produced for the private flyer used types that had been in existence before the war and, in many cases, were "remanufactured" war surplus. In Russia, however, the 5-cylinder, 125-hp M-11D continued to be produced for training planes and for aerobatic competition until well into the sixties. What seems to be the same engine, or one derived

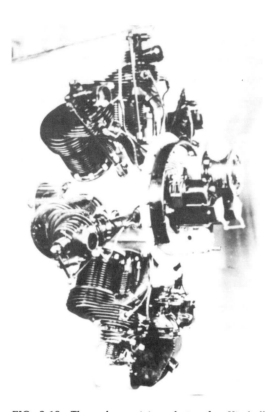

FIG. 8-18 The only surviving photo of a Kimball Beetle radial, 7 cylinders, 135 to 150 hp. Note that in contrast to the usual arrangement of F-head engines, the Kimball had the intake below and the exhaust on top, like the early Renault and RAF types. It is easier to have an intake valve of adequate size when the intake is on top. *(Courtesy of Everett Cassagneres, Ryan Aircraft historian.)*

from it, has been built in Hungary as the M-11FR, at 160 hp, and in Poland as the WN 6 BR 2, at 185 hp. In Soviet use the M-11 series almost always used helmet cowlings, each with its own air shutter, rather than any sort of round cowl. This engine goes back to at least 1926. Korean veterans heard it many times in U-2 trainers that were used as "washing machine Charlie" nuisance night bombers. Evidently it has been a reliable engine and, in their usual way, the Soviet authorities have felt that their engineers have better things to do than spend time developing an engine to replace one that is already doing an adequate job.

In addition, the Poles have developed at least two small radials on their own. First of these was the 7-cylinder WN3 of about 200 hp, first built in 1957 and now called the WSK K-5. From this, apparently, has come the PZL A-1, a 260-hp nine. These engines deliver 194 kW, as rated in the new International System. Since 1979 this new system, which calls for conversion on a basis of 1 hp = 0.746 kW, is supposed to replace the old "steam horses" everywhere.

One small radial has continued to earn its way in America right up to the present day. Continental R-680s, most of them now remanufactured, power the original Grumman/Schweitzer Ag-Cat on spraying jobs all over the country. Sadly, perhaps, most of them are being replaced by Wasp Juniors. Before long they, too, will be gone, victims of the all-conquering opposed engine, and we will hear the evocative growl of the round engine only at air shows and antique meets.

Data for Engines

Maker's name	Date	Cylinders	Configuration	Horsepower	Revolutions per minute	Bore and stroke, in. (mm)	Displacement, cu in. (L)	Weight, lb (kg)	Remarks
Anzani	11	6	R, A	45	1300	3.5 (90) × 4.7 (120)	279 (4.6)	154 (70)	"Total loss" lubrication system
Anzani	22	6	R, A	70–80	1550	4.1 (105) × 4.9 (125)	396 (6.5)	215 (97.7)	
Brownback/Anzani	28	6	R, A	85	1800	4.1 (105) × 4.7 (120)	379 (6.2)	265 (120)	
Potez 6Ac	29	6	R, A	100	NA	4.1 (105) × 4.9 (125)	396 (6.2)	NA	Modernized 70-hp Anzani
Potez 3 B	29	3	R, A	60	2200	4.1 (105) × 4.9 (125)	198 (3.1)	172 (78.2)	
Potez 9Ba	33	9	R, A	250	2400	3.9 (98) × 4.6 (117)	485 (8)	324 (147)	For Coupe Deutsch de le Meurthe race
Salmson AD9	25	9	R, A	45	2000	2.8 (70) × 3.4 (86)	182 (2.98)	150 (68)	
Salmson AC5	27	5	R, A	60	1800	3.9 (100) × 5.1 (130)	312 (5.1)	242 (110)	
Salmson AC 7	27	7	R, A	95	1800	3.9 (100) × 5.1 (130)	437 (7.2)	286 (130)	
Renault 7A	30	7	R, A	120–130	2200	3.9 (100) × 4.7 (120)	403 (6.6)	286 (130)	
Lorraine P 5B	34	5	R, A	125	1650	4.9 (120) × 5.5 (140)	524 (8.6)	354 (161)	
Hispano-Suiza 5Q	29	5	R, A	150	2000	5 (127) × 5.5 (140)	541 (8.7)	407 (185)	
Bristol Lucifer IV	25	3	R, A	130	1700	5.8 (146) × 7.5 (190)	584 (9.6)	325 (147)	
Armstrong-Siddeley Genet	26	5	R, A	82	2200	4 (102) × 4 (102)	251 (4.1)	203 (92)	
Pobjoy "P-type"	28	7	R, A	65	3000	2.4 (72) × 3.4 (87)	151 (2.5)	119 (54)	
Pobjoy "R-type"	31	7	R, A	5	3300	3 (77) × 3.4 (87)	173 (2.8)	135 (61)	
Pobjoy Niagara V	37	7	R, A	125	4000	3.2 (82) × 3.4 (88)	192 (3.1)	175 (80)	

Engine									Notes
Wolseley A.R. 9 (Aries)	27	9	R, A	203	2420	4.2 (106) × 4.8 (121)	589 (9.6)	452 (205)	
Walter NZ60	22	5	R, A	75	1750	4.1 (105) × 4.7 (120)	316 (5.2)	225 (102)	
Walter Polaris I	28	3	R, A	50	1800	4.1 (105) × 4.7 (120)	190 (3.2)	158 (71.2)	
Walter Regulus	28	5	R, A	230	1950	5.7 (145) × 6.3 (160)	697 (11.5)	463 (210)	Published figures given but calculated displacement is 805 cu in.(13.2 L)
Walter Bora II	33	9	R, A	210	2200	4.1 (105) × 4.7 (120)	569 (9.3)	363 (165)	
Siemens Sh 5	23	7	R, A	77	1750	3.9 (100) × 4.7 (120)	403 (6.6)	NA	Rocker arms like Anzani
Siemens Sh 11	25	7	R, A	96	1750	3.9 (100) × 4.7 (120)	403 (6.6)	326 (148)	Z-type rocker arms
Siemens Sh 14	32	7	R, A	113	1720	4.1 (105) × 4.7 (120)	436 (6.6)	308 (140)	Conventional rocker arms
Siemens Sh 14A-4	36	7	R, A	160	2200	4.3 (108) × 4.7 (120)	469 (7.7)	298 (135)	Name became "Bramo" with this engine
Renard Typ. 100	27	5	R, A	145	1750	4.7 (120) × 5.5 (140)	482 (7.9)	275 (125)	
Avia R-7	30	5	R, A	140	2000	4.7 (120) × 4.9 (125)	430 (7)	352 (160)	Skoda prior to 1930
Fiat A-50	26?	7	R, A	95	1800	3.9 (100) × 4.7 (120)	402 (6.6)	273 (124)	
Fiat A-53	31	7	R, A	120	2000	4.1 (105) × 4.7 (120)	443 (7.3)	324 (147)	
Fiat A-54	33	7	R, A	140	2100	4.1 (105) × 4.7 (120)	443 (7.3)	331 (150)	
Farina Algol	28	7	R, A	100	2800	3.4 (87) × 4.3 (110)	279 (4.6)	253 (115)	
Farina T. 58	36	5	R, A	142	1950	4.7 (120) × 5.6 (142)	488 (8.0)	317 (144)	
Alfa-Romeo D 2C 30	35	9	R, A	240	2000	4.7 (120) × 5.3 (135)	838 (13.7)	606 (275)	
Elizalde Dragon V	29	5	R, A	200	2400	4.9 (125) × 5.5 (140)	524 (8.6)	378 (172)	
Tips and Smith Super Rhone	26	9	R, A	125	1450	4.1 (105) × 5.5 (140)	667 (10.9)	315 (143)	Began with Lorraine license
Continental A-70	28	7	R, A	170	2000	4.6 (117) ×	544	415	

(cont.)

Data for Engines (continued)

Maker's name	Date	Cyl-inders	Config-uration	Horse-power	Revolutions per minute	Bore and stroke, in. (mm)	Displacement, cu in. (L)	Weight, lb (kg)	Remarks
Continental R-670	35	7	R, A	210	2000	5.1 (130) × 4.6 (117)	668 (10.9)	430 (195)	
Curtiss Challenger R-600	27	6	R, A	185	2000	5.1 (130) × 4.9 (124)	603 (9.9)	420 (191)	
Lycoming R-680	29	9	R, A	225	2100	4.6 (117) × 4.5 (114)	680 (11.1)	505 (230)	
Wright J-6-7	28	7	R, A	225	2000	5 (127) × 5.5 (140)	756 (12.4)	425 (193)	Called "Whirlwind Seven"
Jacobs L-3	29	3	R, A	55	2125	4.1 (105) × 4.8 (121)	190 (3.1)	189 (86)	One survives and is flying in a Spartan
Jacobs LA-2	32	7	R, A	195	2075	4.7 (121) × 4.7 (121)	589 (9.7)	400 (182)	
Jacobs L-4 (R-755)	34	7	R, A	225	2000	5.3 (133) × 5 (127)	757 (12.4)	415 (189)	
Jacobs L-5	36	7	R, A	285	2000	5.5 (140) × 5 (127)	831 (13.6)	415 (189)	
Detroit Aircat	27	5	R, A	80	2400	4.1 (105) × 3.5 (89)	234 (3.8)	230 (105)	Normal hp was 60
Velie M-5	28	5	R, A	65	1900	4.1 (105) × 3.8 (95)	251 (4.1)	240 (109)	
LeBlond 5D	28	5	R, A	65	1950	4.1 (105) × 3.8 (95)	251 (4.1)	240 (109)	To 70 hp in 1930; iron heads
LeBlond 7D	29	7	R, A	90	1975	4.1 (105) × 3.8 (95)	351 (5.8)	295 (134)	
LeBlond 5DF	30	5	R, A	85	2125	4.3 (108) × 3.8 (95)	266 (4.4)	219 (99.5)	Aluminum (Y-metal) heads
LeBlond 7DF	30	7	R, A	110	2150	4.3 (108) × 3.8 (95)	372 (6.1)	275 (125)	Rearwin (Ken-Royce) after 1937
Lambert R-266	29	5	R, A	90	2375	4.3 (108) × 3.8 (95)	266 (4.4)	224 (102)	
Warner Scarab	29	7	R, A	125	2050	4.3 (108) × 4.3 (108)	372 (6.1)	275 (125)	
Warner Super Scarab 145	32	7	R, A	145	2050	4.6 (117) × 4.3 (108)	499 (8.2)	305 (139)	

Engine								Notes
Warner Super Scarab 165	38	7	R, A	165	2100	4.6 (117) × 4.3 (108)	499 (8.2)	333 (151)	First rating was 75 hp at 1725
Kinner K-5	27	5	R, A	100	1810	4.3 (108) × 5.3 (133)	372 (6.1)	275 (125)	
Kinner B-5	30	5	R, A	125	1925	4.6 (117) × 5.3 (133)	441 (7.3)	295 (134)	
Kinner C-5	31	5	R, A	210	1900	5.6 (143) × 5.8 (146)	715 (11.7)	420 (191)	
Kinner R-5	34	5	R, A	160	1975	4.9 (124) × 5.3 (133)	490 (8.0)	305 (139)	
Szekeley SR-3	28	3	R, A	40	1800	4.1 (105) × 4.8 (121)	191 (3.1)	148 (67)	
Szekeley SR-3L	29	3	R, A	30	1750	4.1 (105) × 4.8 (121)	191 (3.1)	152 (69)	
Szekeley SR-3 O	31	3	R, A	45	1750	4.1 (105) × 4.8 (121)	191 (3.1)	152 (69)	
Aeromarine AR 3	31	3	R, A	55	2400	4.1 (105) × 4 (102)	160 (2.6)	145 (66)	Built later as Lenape Papoose
Axelson A7R	29	7	R, A	150	1800	4.5 (114) × 5.5 (140)	612 (10)	430 (195)	
Security	35	5	R, A	125	1950	4.8 (121) × 5 (127)	441 (200)	355 (161)	
Comet 7 D	30	7	R, A	130	1825	4.5 (114) × 5.5 (140)	612 (10)	NA	
McClatchie Panther	30	7	R, A	150	1900	4.5 (114) × 5.5 (140)	612 (10.0)	439 (200)	
Kimball Beetle	29	7	R, A	135	1800	4.5 (114) × 5.3 (133)	585 (9.6)	380 (173)	Dates and output figures from different sources conflicting
Soviet M-11D	36	5	R, A	125	NA	4.9 (125) × 5.5 (140)	518 (8.5)	344 (156)	
Polish WN 6 BR 2	53	5	R, A	185	NA	NA	NA	NA	
Polish PZL A-1	69	9	R, A	260	2350	4.1 (105) × 5.1 (130)	616 (10.1)	434 (197)	

R—radial; A—air-cooled; NA—not available. Open figures—U.S. Customary; figures in parentheses—metric. Weights are dry.

NINE

The air-cooled inline.

The air-cooled inline engine combines the small frontal area of the water-cooled inline with some of the lightness and simplicity of the radial. These advantages have been recognized since the earliest days of aviation. Glenn Curtiss built the first ones; they were the world's premier lightplane engines between 1926 and 1938; one has held the world's landplane speed record; they powered the planes that trained the RAF, the Luftwaffe, and the Japanese pilots of World War II; and they are still built today (though only in the Soviet bloc).

The first air-cooled inlines to be built in quantity were the early Renault V-8s. As described in Chap. 2, Renault V-8s began at 50 hp in 1910, soon went to 80 hp, were exported to England, were copied there, and saw a certain amount of post-war use. A twelve was added in 1913, first at 90 hp but soon going to 138. All Renault V-types used their timing gears as reduction gears and depended on a blower to assure an adequate flow of cooling air.

The Air-Cooled Inline Reaches Maturity in Britain

The British copy was called the RAF (for Royal Aircraft Factory). The RAF eight (Fig. 9-1) developed 90 hp, and the twelve, in its most-used model, 150. The most fully developed version, the RAF 4D, which had Gibson heads (see Chap. 6), gave 240 hp, but not many were produced. The 150-hp RAF's major application was in the RE (for Reconnaissance Experimental) 8, an observation plane characterized by an odd blend of awkwardness and grace, to which the engine's large air scoop contributed. Although this cooling scoop looked clumsy, it made these the first "stationary" air-cooled engines (if one excludes the early, low-powered Curtisses) to be cooled solely by the airflow generated by the speed of an airplane's passage through the air. As mentioned in connection with the radial engines, the RAF's experiments in cylinder construction culminated in a design that later made the large radial engine practical. The RAF twelves were effective engines for the difficult work

FIG. 9-1   RAF I, 8 cylinders, 90 hp. The aircraft is not identified but is probably a B.E. 2, a reconnaissance type of 1914. A four-blade propeller had to be used on most RAF installations to get sufficient blade area to absorb the engine's power at the geared-down speed of around 800 rpm. *(Courtesy of NASM-Smithsonian.)*

of artillery observation, not least because the absence of a radiator reduced their vulnerability to enemy gunfire.

After the war the British government formed a company called ADC (Aircraft Disposal Corporation) to sell surplus aviation material—especially to the unwary, according to Grover Loening. Among its junk was a large quantity of the big BHP water-cooled inline sixes. In the hope of salvaging something from these deplorable engines, ADC designed air-cooled cylinders for the BHP and put it on the market as the Airsix of 300 hp. This was Britain's first truly indigenous air-cooled inline, and, while few were sold, the experiment led to a long line of successful small engines.

Geoffery de Havilland had built a relatively successful ultralight plane, the de Havilland Hummingbird, powered by a 750 cc (45 cu-in.) motorcycle engine. The Hummingbird's limitations convinced de Havilland that flying with the absolute minimum of power wasn't the answer to anything. Therefore, searching for an adequate lightplane engine, he approached ADC with the suggestion that it build a four out of RAF 4D cylinders. ADC responded with the first Cirrus, which was thus half

of an RAF V-8. Durable and reliable, the Cirrus became the engine that made the Moth possible and laid the foundation for the lightplane movement in Britain.

From this good start, the Cirrus never looked back. When the supply of 4⅛-in.- (105-mm-) bore RAF cylinders ran out, ADC designed a new cylinder of 4 5/16-in. (110-mm) bore, to make the 84-hp Cirrus Mark II [Fig. 9-2(a)]. In 1928 this was followed by the Cirrus Hermes of 94 hp, the first Cirrus to be available in inverted form—ultimately the classic configuration for the air-cooled inline [Fig. 9-2(b)]. In 1935 the 135-hp Cirrus Major and the new, smaller Cirrus Minor of 90 hp were introduced. Two years later the Major was replaced by the New Major of 150 hp and the firm was taken over by Blackburn, an airframe manufacturer that had replaced de Havilland as the most important user of Cirrus engines. The Minor was an outstandingly well-designed motor, but the Cirrus Midget, a 55-hp 1938 introduction, was a rather less prepossessing job with all-iron cylinders.

From 1928 to 1935, the heyday of the air-cooled inline in the United States, Cirrus engines were built under license by the ACE Corporation

FIG. 9-2(a)   Cirrus Mark II, 85 hp. *(Crown Copyright. Science Museum, London.)*

FIG. 9-2(b)   Cirrus Hermes, the first popular inverted air-cooled inline engine. *(Courtesy of Rolls-Royce Limited.)*

of Marysville, Michigan. ACE built the Cirrus in both upright and inverted forms, the latter being called the "Hi-Drive" Cirrus. They were used principally in Fairchild 22s and in Great Lakes Trainers but were not entirely satisfactory in the tough training-school service, the head castings suffering from porosity. ACE gave up after 1935, and the burgeoning Menasco company in California took over the remaining parts and the spares business.

If the Cirrus made the Moth possible, the Moth, in its turn, put de Havilland in the engine business. In 1927 de Havilland had Major Halford, whose reputation had somehow survived his association with the BHP, design a Cirrus replacement for it. The resulting engine was tested in a unique manner. De Havilland built a little low-wing racer, very much like Ben Howard's "Pete" of the following year. De Havilland called it the Tiger Moth and powered it with a 135-hp prototype of its new engine. After demonstrating its ability to withstand the rigors of racing, the engine was given smaller valves and lower compression to derate it to 100 hp, dubbed the Gipsy, and put into production. It was the first of a fine line of powerplants. Probably the greatest thing about the Gipsy, later retroactively named the Gipsy I, was its durability. It was almost certainly the first small engine to be good for a 500-hr TBO. Occasionally it swallowed a valve, but when it did it somehow always kept running until the aircraft found a safe place to alight. It was used in many countries and for many aircraft and was kept in at least nominal production until 1934.

The Gipsy was followed in 1930 by the Gipsy II and Gipsy III, more refined engines in that the valve gear was now enclosed. Both were 120-hp engines, the -III being an inverted version. They came complete with baffles on the cylinders to assure good flow and proper distribution of the cooling air. Many well-known airplanes used these engines, but, to me, the nicest was the de Havilland Puss Moth. The Puss Moth was a two- to three-seat high-wing cabin monoplane with folding wings. Besides having nice lines, it had an interior decor much like that of a good British sports car of the period; inside, one could feel as though one were in a flying version of a Riley Imp or some such. James Mollison crossed the Atlantic in one in 1932,

the first solo east-to-west crossing and the first ever in either direction in a lightplane (Fig. 9-3).

A Gipsy III powered the first Tiger Moth trainers, this being a recycling of the racer's name for a very different airplane. However, the production Tiger Moth that trained most RAF pilots for World War II used the Gipsy Major, the Gipsy III's replacement. The Gipsy Major developed 130 hp; it is worth noting that the British trained their pilots, and trained them well, on only 59 percent of the 220 hp that we employed on the "Stearman."

The Gipsy Six of 185 hp was also introduced in 1932. This engine is memorable for powering two fascinating de Havilland – built twins, the Dragon Rapide and the first Comets. The Dragon series were light transports which, by employing tapered wings of gliderlike slenderness, achieved good aerodynamic efficiency while retaining the biplane configuration. In their day, which has only lately come to its end, their combination of capacity, speed, and economy was a feeder-line operator's dream.

The three Comets were extraordinary airplanes built for an extraordinary event, the MacRobertson England-to-Australia Race of 1934. Comets took first and fourth in the race, though it was a hollow victory considering that second and third places were taken by a Boeing 247 and a DC-2. The Gipsy Sixes used were special racing versions intended to give above-normal fuel economy as well as extra power.

In 1937 de Havilland built on its Gipsy Six experience to move up to a higher horsepower class with the Gipsy Twelve. While the Gipsy Twelve used cylinder assemblies from the Six, it gave 525 hp as against 200 for the Six—more than twice the power. It also weighed more than twice as much. Much of the power gain and some of the weight increase came from the addition of a supercharger, de Havilland's first. Nevertheless this is probably the only case of a closely related six and twelve where the twelve weighed more than twice as much as the six.

The Gipsy Twelve makes an interesting comparison with an earlier British air-cooled twelve, the RAF4D of just 20 years previous. The RAF developed 240 hp, 46 percent of the Gipsy's power, from 806 cu in. (13.2 L), 85 percent of the Gipsy's displacement. This comes to 3.35 cu in./hp for the RAF and to 2.14 cu in./hp for the de Havilland (18.2 hp/L and 28.5 hp/L, respectively), a much smaller improvement than was common over the period for other types of engines. An obvious example of this is the fact that the 1917 Liberty gave only 40 percent as much power per unit of displacement as the 1937 Allison of comparable size. Of course, the RAF4D was the culmination of considerable development work whereas the de Havilland was the first of its line, even allowing for its makers' previous experience with the six. Had demand for the Twelve warranted the effort, more power probably could have been had from the Gipsy Twelve's basic design. However, no one but de Havilland's own airframe division used the engine, and no further development took place.

The difficulty with an air-cooled V-12 is that it operates in a power range where the advantages of an air-cooled inline are of little real significance. That is, an engine of 300 to 600 hp would ordinarily be used in an airplane whose fuselage was wide enough to seat two people side by side, say 40 in. (1016 mm) or more. For such planes a radial is hardly bulkier than the basic fuselage, and the resulting drag is no more than what it would be with an inline. In these applications, the radial's lighter weight and longitudinal compactness would probably make a designer decide in its favor. Airplanes like the Fairchild 24 illustrate the fact that the use of a radial did not harm performance in bulky-fuselage airplanes. This three-place (four-place in later models) high-wing cabin plane was available with either a Warner radial or a Ranger inverted inline six, both of 145 hp. Top speed was the same[1] with either engine, but with the Warner the empty weight was 112 lb (51 kg) less and, somehow, climb was 28 percent better at the same gross weight. What the Ranger did do was transform an ordinary-looking airplane into a beautiful one.

Britain had one other significant maker of air-cooled inlines: Napier. Major Halford designed the most remarkable of all air-cooled engines for

[1]Nominally, with full load—a higher useful load for the 24W. As Ev Cassagneres has pointed out, the *same* load in both planes results in a lower actual weight for the 24W with a resulting higher speed.

Napier in 1930. Called the Rapier, it was a 300-hp H-16. That is, it had four rows of 4 cylinders each, but these did not extend outward from a single crankshaft as they would have in an "X-type." Instead, there were two crankshafts, one on each side, geared to a central propeller shaft. Each crankshaft was driven by 4 upright cylinders and 4 inverted ones. Each pair of banks shared a common camshaft which operated the valves through short pushrods. The point of the whole confection lay in its compact dimensions: 35½ in. high × 21 in. wide × 51 in. long (890 mm × 533 mm × 1295 mm) is not bad for a 300-hp engine. It developed its rated power at 3500 rpm, which must have produced a Ferrari-like exhaust sound from 16 cylinders.

This startling engine does not seem to have been used in any production aircraft, nor does its offshoot, the 150-hp 6-cylinder E 97; but another offspring, the Dagger, saw limited service. The Dagger was a 24-cylinder outgrowth of the Rapier, introduced at 805 hp. It first flew in the Hawker Hector Army cooperation biplane. The Hector was a reengined Audax, which was really a modified Hart. Honest! By World War II the Dagger, now up to 1000 hp, had found its way into a Handley-Page Hampden variant called the Hereford. While reliability was not of the highest order, the main reason for its being dropped from service was its effect on aircrew health and morale. With 48 cylinders blatting away at 3500 rpm, the crews were deaf for several hours after every mission. The performance of the Hereford was no better than that of the Pegasus-powered Hampden, even though the frontal area of the engine nacelles was noticeably less.

France and the Lightweight Fighter

The lightweight fighter was another experiment with military applications of large air-cooled V-12s. Both France and the United States looked into the possibility of building effective baby fighters around air-cooled inlines, the French more seriously than the Americans.

The French lightweight fighter was a natural outgrowth of three concepts. One was a class of

FIG. 9-3(*a*)   de Havilland's first engine, the 100-hp Gipsy I.

FIG. 9-3(*b*) The de Havilland DH 60 Gipsy Moth, the world's preeminent
lightplane of the twenties and early thirties. Modified with steel tube (instead of
plywood) fuselage, swept-back wings, and an inverted engine, it became the
Tiger Moth, Britain's primary wartime trainer.

FIG. 9-3(c)   The de Havilland Puss Moth in which Captain James Mollison flew the Atlantic. The name came from a Newfoundland village that was to have been his destination. Expensive to build, it was replaced by the much less attractive Leopard Moth.

single-seaters, used around 1930, that were called "jockey" fighters. They were designed around a power limitation of 450-hp maximum and may have taken their name from a restriction on pilot weight. Another element was the remarkable series of Caudron racers of the mid-thirties; and the third was the development by Renault of a suitable inverted twelve.

Renault had reentered the air-cooled inline field in 1928 with its 4 Pb, an upright four devel-

FIG. 9-3(d)   de Havilland Gipsy Twelve. In the de Havilland Flamingo transport, the Gipsy Twelve was cooled from the rear. Air was taken in at openings in the wings and ducted forward. (Courtesy of Rolls-Royce Limited.)

oping 90 hp at 2000 rpm. This was very much the counterpart of the early Cirrus and Gipsy uprights, having a wet sump and exposed valve gear that had to be lubricated with a Tecalemite grease gun before each day's flying. Between 1928 and 1931 this little workhorse was the dominant French engine in its power class. Its upright configuration gave rise to some front ends that rivaled in ugliness anything the British ever did with two-row radials.

In 1932 the inverted 4 Pci, a rather more advanced motor, was added. It had enclosed valve gear and developed 120 hp from the same 355-cu-in. (5.8-L) displacement as its predecessor. A year later the 4 Pci was enlarged to 386 cu in. (6.3 L) as the 140-hp 4 Pdi, or Bengali. This was a very fine powerplant; even the sometimes absurdly chauvinistic editor[2] of *Jane's All the World's Aircraft* called it "one of the world's great engines."

Renault built its first sixes in 1934. The standard model was the 6 Pdi, using Bengali cylinders and delivering 180 hp. The Special was quite different. With smaller-bore cylinders to bring its displacement down to the Coupe Deutsch de la Meurthe limit of 8L. The Special displaced 485 cu in. (7.95 L)—and with a large supercharger, it gave 325 hp at 2350 rpm. In Caudron racers, the Special won the Coupe Deutsch, took the world's land-plane speed record away from an 800-hp Weddell-Williams, and came to America to win the Thompson Trophy over planes powered by radials of more than twice its power.

When Renault came out with its twelve, it was natural to think of using it in a military airframe based on the advanced design of the Caudron racers. Renault's twelve, the 12R. 01, was in one sense more conservatively designed than its de Havilland counterpart. From a displacement slightly greater than the Gipsy Twelve's, it produced less power—450 hp to the de Havilland's 525. However, as its weight was only 80 percent of the Gipsy's, its power-to-weight ratio was marginally better.

At 302 mph (486 km/hr), the 1939 Caudron

[2]Charles Grey Grey. Mr. Grey, for instance, habitually handled the superiority of American civil aircraft over British by denying—not questioning—American performance test data.

C 714 fighter designed around this engine was as fast as many fighters of twice its power and, with four wing-mounted guns, had comparable firepower. This performance was achieved by using a small wing of 29-ft (7.92-m) span, and so its climb suffered and its landing speed was high. Badly as France needed anything that could fly and shoot, fewer than 100 were built and many of these went to Finland.

After the Renaults, the most important French air-cooled inline engines were the Regniers. These began as French licensees for the Gipsy line, Regnier's R. 4 of 120 hp and R. 6 of 210 being essentially the same as the corresponding de Havillands. Regniers, like most license-built engines, soon diverged from their origins. First the four was dropped, then the six became the 6B-03 with some homegrown components. Then the six was made available in supercharged form, giving, at 280 hp, a 51 percent power increase over the original Gipsy Six. In 1936 two fours of indigenous design were introduced, the 60-hp 4 J and the 90-hp 4 L. Regnier model designations are hard to unscramble, but there were also 70- and 145-hp fours and 230- and 250-hp sixes (Fig. 9-4).

In 1936 Regnier took the same step as many others by making a V-12. It was a geared engine with sufficient offset to the propeller shaft to permit firing a small cannon through it. Like the Renault twelve, it was a 450-hp engine, but, unlike the 12 R. 01, it never got past the experimental stage.

Until the fall of France, Renault 100-hp 4 P-03s were produced to power the Stampe trainer but Regnier production was apparently halted. Post-war, however, few if any Renaults were built, but the Regnier line was revived. When the government took over the Renault business, it dropped the aviation line and concentrated on cars. Regnier was also nationalized as part of SNECMA, an aircraft engine group, and production continued until the general acceptance of opposed engines ended the demand for inverted inlines.

The Renaults and Regniers were not France's only air-cooled inlines. The Chaise, unique in being an inverted V-4, was fairly well known. First made in 1930 at 100/120 hp, the 1932 model gave 130 hp, and the 1934 gave 140. In addition to a narrow angle between banks, reminiscent of the Lancia V-

FIG. 9-4   Post-war SNECMA-Regnier 4L-04, 100 hp. Essentially the same as the pre-war, 90-hp 4L. Except for a somewhat long nose, the Regniers were typical European inverted air-cooled fours. (*Courtesy of NASM-Smithsonian.*)

4 automobile engines, Chaise used four unusual design features. Its first engines had iron cylinders cast in one piece with the lower half of the crankcase. This is really a very fine way to build a small engine if the available foundry skills are up to the difficult job of making such complex castings. It combines lightness, strength, and economy, and one knows that the cylinders will never come loose or leak at the base. Evidently, though, Chaise's foundry was unable to cope with this demanding design; it was used only on its first model, the AV 2. The two succeeding engines reverted to conventional separate-cylinder construction.

The second odd feature was the use of two camshafts placed outside the vee of the cylinders. The oddity lies in the fact that one of the attractions of the V-type layout is its ability to employ a single camshaft inside the vee, between the banks. Also unusual was the employment of ball bearings, necessitating the use of a crankshaft built up in sections instead of the more usual one-piece shaft. Finally, the rocker arms, valve springs, and push-rods were not enclosed in oil-tight covers, the Chaises being among the very few inverted engines so designed. Since the line ended with the 1934 models, it may have been a case of "For many an innovation, One pays in botheration."

The Michel Am 14s constituted another minor French line. These were upright, exposed-valve-gear wet-sump fours of 90, 122, and 130 hp. Made in Strasbourg by a firm said to have also made cars and motorboats,[3] they were on the market from about 1928 until 1934.

Train, of Courbevoie, began a year after Michel gave up. Train pursued the idea of using the same cylinder in both a four and a six to its logical conclusion by building a two as well. This gave it a 20 hp, a 40 hp, and a 60 hp. Train was one of a number of European motorcycle-engine builders who dabbled in minimum-size aircraft engines; before motorcycle engines ran at the astronomical speeds so common today, there was a certain degree of affinity between the two types of powerplants. The firm's products found their way into a number of ultralight aircraft, including pre-war Pous de Ciel and post-war Jodels.

The Air-Cooled Inline in America

Perhaps American thinking was conditioned by the fact that the first U.S. inline air-cooled engines were well above what was then the lightplane power

---

[3]I cannot find the Michel car in any automotive history, but this is not conclusive proof that none were built. For instance, I have seen an English car called the Boyd-Aurea, but nobody else that I know has ever heard of it.

range. Whether for this or for other reasons, the air-cooled inline always took a secondary place in the United States. American preference lay with radials at first and later with opposed engines.

The first American air-cooled inline of the twenties bore the surprising name of Allison. It was an air-cooled version of the Liberty and was called, after its displacement, the V-1410—one of the first engines to use the identification system that eventually came into general use in the United States. The displacement of the air-cooled Liberty was 14.5 percent less than that of the water-cooled version, something that serves to illuminate a fact of some importance in the design of air-cooled inlines. The cooling fins of an air-cooled cylinder take up more space than the water jacket of a liquid-cooled one. This means that either the cylinders must be spaced further apart if the bore is the same or the bore must be reduced if the spacing is unchanged.

With Allison's air-cooled cylinders for the Liberty base, the smaller displacement resulted from the necessary reduction in cylinder bore. Yet the horsepower went up, not down. At 420 hp, the V-1410's output was the same as the late high-compression Liberty's and more than that of the standard model. As noted in Chap. 1, the ability of an engine to produce power is largely a function of its breathing and its compression ratio, and both of these had been enhanced in the V-1410. But there is also the question of the engine's ability to stand up to the power it produces. This will depend on the heat dissipating capacity of the cylinder and head and on the ability of the lower end to take the bearing loads generated. In the case of the Liberty, the lower end had shown that it could stand up to an output of 35 hp per cylinder, and so if a smaller cylinder could generate the same power as the original one, success was purely a matter of heat dissipation. As Fig. 9-5 shows, Allison's engineers had learned the Gibson-Heron lessons well, and, given proper cowling and ducting, the V-1410 cooled very well.

The air-cooled Liberty was apparently a good engine. It passed all its tests and gave a D.H. 4 improved performance when one was tried in place of the standard water-cooled Liberty. However, this was not enough to lead to a reversal of official policy. The Army continued to favor the D-12, and

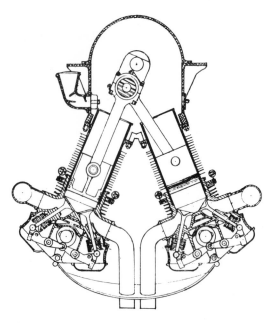

FIG. 9-5   Cross section of Allison V-1410 (inverted air-cooled Liberty). The arced lines at the bottom represent the airscoop, very necessary at early-twenties speeds. (*Courtesy of Detroit Diesel Allison Division, General Motors Corporation.*)

the Navy the radial. Allison's engine was heavier and bulkier than the D-12, and so it could not have replaced the Curtiss engine in fighters, but one would think that it might have given the Army's heavy bombers a few more years of useful life had it been used as a Liberty replacement.

Curtiss had a try at the same sort of thing in 1929 or 1930 with its V-1460. As the Allison was to the Liberty, so, roughly, was the V-1460 to the Conqueror. As with the Allison, the Curtiss V-1460 performed well on trials but got more publicity than production.

The first civilian U.S. air-cooled engine was the 125-hp Dayton Bear of 1927. It was a product of the same line of thought as the Allison and the Airsix, in this case the Hall-Scott L-4 forming the basis for the conversion. It was light, at 375 lb (170 kg), for an engine that developed its power at an efficient 1800 rpm, and its fuel consumption was low at 0.505 lb/(hp)(hr) [0.229 kg/(hp)(hr)], but it met with little success in the market. Who wouldn't rather have had one of these dual-ignition engines than an OX-5 in the nose of a Swallow or Eaglerock? Perhaps the vibration or the cost, or both, stood in its way. Another firm tried something similar on a smaller scale in 1930 with the Dayton Aero Four. This involved using Model A Ford internals with overhead-valve air-cooled cylinders. Again, there were few if any takers.

In 1928 the above-mentioned American Cirrus appeared, to become the first commercially successful U.S. air-cooled inline. Within the year it was followed by a spate of other makes. Of these, the longest lasting was the Fairchild, first introduced as the 6-370. It was a fairly sophisticated engine, and while it was considerably enlarged during its 17-year production life, its basic design underwent very few changes. It had less displacement than many fours, making for a smooth engine, and, at 325 lb (147 kg), was reasonably light for its 122 hp.[4] In 1930 Fairchild changed the name of its engine division to Ranger and enlarged the six to 390 cu in. (6.39 L) and 150 hp. From 1935 on a slightly detuned version, the 145-hp 6-390D, was offered, and, in 1939, a 410-cu-in. (6.7-L) model developing 165 hp joined the line. In 1941 it was again enlarged, this time to 440 cu in. (7.2 L) and 175 hp. Wartime versions included the I-440-1 of 170 hp, the I-440-3 of 200, and the I-440-7 also of 200 (Fig. 9-6).

Fairchild/Ranger engaged in some careful experiments with airflow, resulting in the finding that air passing more than 3/16 in. (4.76 mm) away from the cylinders contributed nothing to cooling. On the basis of this knowledge, Rangers came equipped with close-fitting baffles around the cylinders and were able to use smaller air-intake openings in their cowlings than most similar engines could tolerate. The Ranger sixes were reliable but quite dirty, a condition that may have been exacerbated by the absence of excess airflow.

Ranger sixes were used almost exclusively in Fairchild's own planes, the aforementioned Fair-

FIG. 9-6   Early Ranger six, 150 hp. The intake manifold design looks wrong; it would seem that the center cylinders, nos. 3 and 4, would be somewhat starved for fuel compared to nos. 1, 2, 5, and 6. All Rangers, both sixes and twelves, had the same characteristic crankcase appearance. *(Courtesy of the author, taken at Glenn Curtiss Museum, Hammondsport, New York.)*

child 24 and the wartime PT-26. However, as over 6600 of these were built, total Ranger production was greater than that of many engines used in a wider variety of aircraft.

Ranger introduced its twelve, the V-770 (12.6 L), in 1931. It was little more than a paper engine for several years. Fairchild may have flown it in an experimental installation or two, but I know of no production airplane that used it. Power crept up as time passed: 300 hp at first, 350 later in the same year, 420 in 1932, and 520 in 1940. It powered America's only lightweight fighter, the Bell XP-44. Its design parameters were similar to the Caudron's, but it differed in having a bubble canopy and in using a tricycle gear, presumably because of having a high landing speed.

The only plane that used the V-770 in production was the Curtiss SOC-3 Seagull, one of America's worst aircraft of the Second World War.[5]

[4]But notice the weight penalty involved in using an increased number of cylinders. The weight and power are very close to those of the Dayton Bear, of only 4 cylinders; the Fairchild, being a design of 12 years after the Hall-Scott, would have been expected to be lighter.

[5]When the definitive title of "World War II's Worst" is finally awarded, the aircraft will probably carry a Curtiss-Wright nameplate. While the F2B and P-39 were bad, they started good and were ruined by government-mandated changes. Planes like the Seagull and Helldiver began bad and were painfully cobbled into marginally usable jobs. Curtiss could produce very well but not design.

This was a shipborne observation floatplane which replaced, and whose deficiencies were emphasized by comparison with, the classic Chance Vought OS2-U Kingfisher. When flown slowly, as many observation missions required, the engine tended to overheat and quit without warning. When it did, the Seagull would dive some distance before building up enough speed to come back under control, making for a messy situation when the incident took place at low altitude (Fig. 9-7).

The Chevrolair was another 1929 design. It had no connection with the Chevrolet car, though some writers have assumed that it was a conversion of the automobile engine. The name came from Arthur Chevrolet, its designer. It was an advanced design for its time, using double overhead cams to allow cooling fins in the angle between the valves, much like those on the Allison or on a radial. Unfortunately, this was largely wasted effort since it was based on the assumption that the cooling air should flow from front to back. As Fairchild and others were showing at about the same time, an inline is best cooled by bringing the air in off-center

and passing it across the cylinders. Every air-cooled inline installation designed after 1930 shows this by its off-center air-inlet opening.

There were two Chevrolairs, a 120-hp four and a 180-hp six. They found some sales for such little-known planes as the Doyle Parasol, but the firm gave up in 1932. It was then absorbed by Glenn L. Martin, who built it for 2 years as the Martin-Chevrolet. In 1939 the design was revived by a company called Phillips, which was building the former Driggs Skylark under its own name and which had developed an attractive low-wing all-wood trainer, both of which could make use of the ex-Chevrolair engines. Phillips also tried to develop an inverted V-8 based on the Chevrolair four, but the war ended its efforts.

The Michigan Rover was a somewhat smaller 1929 production. It was an inverted four of sound, conventional design with pushrod-operated valves, getting 55/60 hp from 236 cu in. (3.87 L). In 1930 the displacement was raised to 267 cu in. (4.4 L) and the output to 75 hp. The year 1933 was the firm's last in production, but in the meantime the

FIG. 9-7 Curtiss SO3C Seagull in landplane form. The turned-up wingtips and drooped airscoop were afterthoughts not present in the prototype. They didn't help much. As someone said in another connection, no airplane that looks like that can possibly be any good. (Courtesy of NASM-Smithsonian.)

Rover had seen service as the standard engine for the Driggs Skylark and as the original powerplant for the "Queen of Parasols," the Fairchild 22—two of the nicest light open-cockpit two-seaters ever made in America.

Another 1929 inline resulted from Curtiss-Wright being granted a license to build the Gipsy line. This did not have the impact on the American market that might have been expected, for two reasons. One was that Curtiss-Wright had no aircraft in its line that could have used the Gipsy, and so there was no assured sales base. The other problem was that Curtiss-Wright's efforts in the engine field at the time were suffering both from lack of direction and from problems involving the consolidation of the former Wright line with that of Curtiss. When the firm decided to concentrate on large radials, the Gipsy was dropped and the existing stock of spare parts was turned over to Menasco in Los Angeles.

There were two other American air-cooled inlines of around 1928 that can be considered, to a degree, as oddities. Best known of these was the Heath, or Heath-Henderson, B-4. Ed Heath's once-famous Heath Parasol, an ultralight single-seater, had for some time been an airplane in search of an engine. To solve this problem, Heath developed an aircraft conversion of the Henderson 83-cu.-in. (1.36-L) 4-cylinder inline motorcycle engine. It had cast-iron cylinders and heads, and the rear cylinder tended, in the absence of effective baffling, to run hot; therefore Heath simply lowered the compression ratio of the number four cylinder. Apparently this worked well enough in the undemanding ambience of pre-1930 sport flying. It is easy for us to forget how little performance it took to satisfy the impecunious enthusiast in the twenties. Not long ago someone built an LN, last and best of the Heath Parasols, with Volkswagen power. The takeoff and climb were so dismal by modern standards that the VW was replaced with a 65-hp Continental, yet the VW conversion had been a more effective engine than the prototype's Heath-Henderson (Fig. 9-8).

The "Tank," officially the Milwaukee Tank Company V-502, was another somewhat strange engine. It was a 115-hp air-cooled conversion of the venerable OX-5. Since the lower end of the

FIG. 9-8 Two Henderson conversions: a Heath B-4 on the left and a private individual's home-made conversion on the right. Note that propeller rotation is opposite, the Heath having "European" rotation. The propeller on the homebuilt conversion is patterned after the old Reed, first metal prop to see wide use. (*Courtesy of the author, taken at Glenn Curtiss Museum, Hammondsport, New York.*)

OX-5 was far better than the cylinders, ignition, and cooling system, the idea had its points; it was just several years too late. The Tank was used in a few Curtiss Robins, and a "Tank Robin" was a nice airplane, but it had hard sledding against the low Depression prices of the new small radials. It dropped out of the market after only 3 years. The same firm also designed a new air-cooled V-8 of smaller displacement and the same power, but this, too, failed to make a significant impact on the market.

The last of the American inline lightplane engines came out in 1940, just in time for the war to kill any hopes of success that it might have had. It was called the Skymotor and had an underhead camshaft with the unique and, it would seem, sensible feature of an oil sump incorporated in the camshaft cover. It gave 63 hp from 201 cu in. (3.29 L) and advertised "94 fewer parts"—than what, was not stated. All innocent enough—but it was made in Milwaukee, and many of its internal parts were from the Model A Ford automobile engine![6] Was this a last try by the old Tank people (Fig. 9-9)?

[6]Why not? Even in water-cooled form, a Model A could be converted into a fairly decent airplane engine (see Chap. 11).

New ADVANCED ENGINEERING in SKYMOTOR ... for PERSONAL PLANES

This radically different power plant makes ownership of small planes more attractive than ever before. SKYMOTOR design insures Safety, Efficiency and Dependability. Smooth operation eliminates flying fatigue. Further facts mailed on request.

★ FULL VISION
MORE STABILITY
LIGHT WEIGHT
GREATER ECONOMY
94 FEWER PARTS
PRECISION BUILT
63 H.P.—2150 R.P.M.
A.T.C. NO. 200 ★

For Descriptive Folder . . . write to **SKYMOTORS, INC.**
Aircraft Engine Mfrs., 717 So. Sixth St., Milwaukee, Wis.

FIG. 9-9 Skymotor, the last American inverted air-cooled inline, 1941. Well-conceived but it was the wrong motor at the wrong time. Just how much Model A Ford was in it is uncertain. It had seven main bearings and the power suggests bigger valves than the Model A.

Almost as long lasting as the Ranger was the Menasco. Menasco had entered the engine business by making an air-cooled conversion of the old water-cooled wartime Salmson radial. Then, in 1929, it came out with its 4-A Pirate, an inverted four giving 90 hp at 1950 rpm. In 1930 it became the B-4 and power rose to 95 hp at 2000 rpm. Also in 1930, the C-4 was offered; this was the same basic engine but was bored out ¼ in. (6.35 mm) to raise the displacement from 326 cu in. (5.34 L) to 363 cu in. (5.95 L) and its output to 125 hp at 2175 rpm (Fig. 9-10).

To round out what must have been a busy year, Menasco also introduced its first six, the B-6 Buccaneer, of 160 hp. While little used in production planes, the Buccaneer became a racer's dream,

and it was eventually cataloged with outputs as high as 315 hp, from a mere 415 lb (189 kg).

All Menascos had pushrod-operated valves set crosswise on the head and made to operate off a single camshaft in the block by the simple expedient of allowing the exhaust pushrod to carry a pronounced slant. The only other unconventional aspect of Menasco design was the use of exposed valve gear, a retrograde element that persisted as late as 1940.

Menasco engines had a split personality. In stock form, as used principally in the Ryan S-T, a Menasco would rumble along forever at its comfortably low rpm, but the racing Menascos were temperamental screamers. These completely dominated their displacement classes in American racing during the thirties. "Al" Menasco, the firm's president, took a personal interest in building souped-up engines to order, and factory cooperation was excellent. As with most racing engines for any medium, running a hot Menasco called for incantations and spells as well as more mundane forms of expertise. For example, the special 544-cu-in. (8.91-L) 6CS-4 Super Buccaneer that powered the Schoenfeldt "Firecracker" boasted the following after-delivery modifications: modified cooling baffles; high-compression pistons; a special camshaft;

FIG. 9-10 Menasco C-4 Pirate, Los Angeles-built, 1930 to 1941. Power plant of the much-admired Ryan S-T. (*Courtesy of NASM-Smithsonian.*)

oversize valves; higher-ratio gears for the super-charger; new manifolds; air ducting to cool the magnetos; a skin radiator to cool the oil; and the development of a special fuel mixture. With all these changes, the engine would run well at one speed only—3500 rpm. So operated, it was good for 10 hr operation before needing an overhaul.

Menascos were used in one experiment that belongs in any compendium of aeronautical oddities. Alan Loughead, in retirement after selling his interest in Lockheed Aircraft to Detroit Aircraft Corp.,[7] modified a Lockheed Vega to try a novel engine installation. Called the Alcor, it had two 130-hp supercharged Buccaneers set flat, head to head, with the crankshafts 93 in. (2360 mm) apart. It was, in other words, a twin-engine airplane with both engines in the fuselage. It had only one advantage: control with one engine out was very good. Unfortunately, performance was lower than that of a standard Vega having 10 fewer horses and 40 lb (18 kg) more weight. It was just a little odd-looking; its fuselage resembled a hammerhead shark.

The Inline Air-Cooled Engine in Germany: Argus and Hirth

Two German firms, Argus and Hirth, built fine air-cooled inlines during the heyday of the type. Argus will be remembered as a very early aircraft-engine maker who had dropped out of the business after the First World War. It came back into the field in 1928 with a quickly forgotten water-cooled V-12 and an inverted air-cooled four. The four was a 100-hp pushrod engine of 386 cu in. (6.3 L), called the As. 8. Unlike most similar engines, it had the magnetos at the front, a long forward extension of the crankcase and shaft making this possible. It would seem that this feature would contribute to cool magnetos and a well-streamlined nose shape,

but no other manufacturer used it (except Pratt and Whitney, but the Wasp Major was a somewhat different engine).

In 1931, Argus built one of the very few inverted air-cooled V-8s to ever reach production status, using As. 8 cylinders. Called the As. 10, it produced 220 hp from 771 cu in. (12.6 L) at 2000 rpm. This is actually a slightly lower specific output than that of the water-cooled 220-hp Hispano-Suiza of 1918, but it was achieved at a weight of only 445 lb (202 kg) complete. If one of these engines were substituted for a direct-drive modern opposed six in one of today's planes, the Argus's lower revs would lead to higher performance and lower fuel consumption. This is more than a theoretical analysis. I have heard of a Messerschmitt Me. 108 "Taifun" in which a flat six of higher power replaced the original 240-hp Argus As. 10C. Performance fell off, and fuel consumption went up. As a bonus, visibility over the nose went from poor to terrible.

The Germans used these Argus V-8s in single-seat high performance trainers, a type of plane that I don't think anyone else built. Two in particular, the Arado AR176 and the Focke-Wulf FW-56A "Stosser" (Fig. 9-11), were as strikingly attractive a pair of planes as can be imagined, and the exotic shape of the inverted V-8's cowling contributed greatly to their looks. The fours, competing with the extraordinary Hirths, were used less. Argus also cataloged a 230-hp six in 1933 and a 450-hp twelve in 1935, but it is doubtful that significant quantities were made. For the most part, an Argus was a V-8.

And now the Hirths. Once in a great while a piece of work is encountered that gives the impression that its builder worked on a basis of doing everything right with no concessions to custom or to economics. The Hirth engines had this perfectionistic character. No plain bearings were used; the mains were ball bearings, and the connecting rod and wrist pins had rollers.[8] This meant that the crankshaft had to be taken apart to assem-

---

[7]"Lockheed" is Loughead phoneticized. Detroit Aircraft, incidentally, promptly ran Lockheed into bankruptcy: a typical case of what happened when business executives tried to apply their methods in a field that required the touch of knowledgeable enthusiasts rather than good accounting and the rest of the business executive mystique.

[8]It is almost impossible to use ball bearings for connecting rods. "Brinelling"—marks in the races from the balls being pounded in by the explosions in the cylinders—occurs and destroys them.

FIG. 9-11 Focke-Wulf FW-56A "Stosser" advanced trainer of the mid-thirties. Take one glider pilot, add time in a "Stosser" and the result is apt to be an excellent fighter pilot. *(Courtesy of the National Archives.)*

ble or disassemble the engine, and so Hirth invented the best method yet for locking the sections of the shaft together. With all bearings being anti-friction types, there was no point in having oil in the crankcase, but this meant that the internal cooling effect of the ordinary oil-circulating system was absent. To make up for this, filtered fresh air was passed through the crankcase. It might be noted that Hirth was many years ahead of its time in this respect. Designers of large stationary Diesels are now isolating the cylinders from the bearings instead of having them share a common lubricating system.

On the Hirth engines, the air from the crankcase passed into the carburetor. Since it was warm, the possibility of carburetor icing was largely eliminated (though at some cost in reduced power). This method automatically regulated the internal cooling, since both the heat generated and the volume of air passing through the crankcase were proportional to the throttle opening.

Hirth's rigorous approach to design also produced a unique solution to the problem of designing a dustproof and oilproof cover for the pushrods. Of the four long studs that held down each cylinder, two were elegantly necked down, and the other two were hollow with the pushrods inside. A short camshaft ran across the engine between every other pair of crank webs, and so the hollow studs were at the rear of the number one cylinder, the front of number two, and so on. Nothing was overlooked that could possibly make for a better engine. The pistons were apparently imported from America at first. To promote easy starting, coil ignition—dual, *naturlich*—was used. The crankcase was magnesium (then a much more exotic metal than now), not the more usual aluminum.

None of this design elegance was a matter of an academic engineer with a big budget going off the deep end (automotive enthusiasts will recollect the excesses committed by Wilfredo Ricart on the Pegaso), and it all worked to make a durable, trouble-free engine.[9] The Hirths, especially Helmut, were all practical aviation people, and their engine was an eminently practical one, in all probability the best of all air-cooled engines of its size.

Hirth made some biggish engines, but its main fame rests on the small ones. Its first was the 1931 H-M 60, a four of 211 cu in. (3.46 L), 65 hp at 2100 rpm, and 181 lb (82.3 kg). The 1933 H-M 60R was slightly bored out to 219 cu in. (3.59 L), and with higher compression gave 80 hp at 2400 rpm. Other Hirths included the 90-hp H-M 4, the V-8 H-M 150, apparently not produced in series,

the 135-hp H-M 6 of 1940, and two other abortive essays into higher power, the 245-hp H-M 8 and the 450-hp H-M 12. Hirth engines powered the first Bucker Jungmeister, elegant low-wing monoplanes made by Klemm and Feisler, and that queen of two-place sport biplanes, the Bucker Jungmann.

Helmut Hirth died of kidney disease in 1939, and the firm passed into Ernest Hienkel's hands, to be used as a development shop for jet engines. When private flying revived in post-war Germany, the costly design features of the Hirth engines could not justify themselves in competition with mass-produced American opposed engines. When Hirth went bankrupt in the mid-seventies, its main products had for some time been two-stroke engines for snowmobiles and "self-launching" sailplanes. Jim Bede's little BD-5 used a Hirth engine, and the Hirth firm's bankruptcy was one Bede problem that wasn't self-created. Hirth engines were made in only one country other than Germany: Hitachi-built Hirths powered many, if not most, of the Japanese military trainers.

Czechoslovakia and Italy

In Czechoslovakia there were two builders of air-cooled inlines, Walter and Zlin. Zlin was a minor builder who made only one model before the Nazi takeover, but Walter is today the main surviving producer of the type. The Walters have been quite conventional but excellent engines. The first Walter inline was the Junior of 1932. Since it was a 105- to 120-hp engine from a firm then making radials as small as 50 hp, one wonders what it was junior to; but never mind. In 1933 it made a bored-out version developing up to 130 hp and called it the Junior Major, possibly to show that the British had no monopoly on silly names. The Minor, which quickly became a very popular engine, and the Major Six joined the line in 1934. Then, in 1935, the Mikron, the smallest well-known four, at 133 cu in. (2.18 L), was added. This 56-hp engine also enjoyed considerable popularity (Fig. 9-12).

Walter also made twelves. One, the Minor Twelve, was apparently a racing engine. It gave 400 hp at the high speed of 2750 rpm, with the aid of a supercharger, and came with 3 to 2 gearing

---

[9]Though mechanical noise, as on any ball-and-roller bearing engine, was a bit high. Apparently, it sort of growled as it ran.

FIG. 9-12 **Walter Major six, 190 to 205 hp. A development of this very good engine is the last surviving production air-cooled inline.** *(Courtesy of the National Archives.)*

and attachments for a variable-pitch propeller. The other twelve was the well-known Saggita, one of the largest air-cooled inlines of recent times at 1120 cu in. (18.35 L). Its size did not result in correspondingly high output; the Saggita developed only 400/420 hp. It had the same gear ratio as the Minor Twelve. The Minor Twelve was introduced in 1938, the Saggita in 1936.

Walter engines were used in Belgian, Italian, and British, as well as Czech, aircraft. Their most memorable achievement has been their powering of the Zlin "Akrobats" that have won several postwar aerobatic championships. For such use, the Walters are equipped with inverted fuel and oiling systems and seem to run just as happily upside down as they do when upright. As the Zlins roll onto their backs there is a momentary hesitation, known as the "Zlin cough," before they settle down to inverted operation, but this did not seem to bother the Communist-bloc pilots who flew them so skillfully.

In 1970 the Walter engines in the competition Zlins were replaced by 210-hp Avias. The name is a little misleading. There had been a Czech

firm called Avia, whose Polish branch had built a small inverted four from 1936 up to the German invasion, but the Czech plant made only opposed engines. What has happened is that the government has consolidated all aircraft piston engine manufacturing under the Avia name; but the Zlin's powerplant is still basically the Walter design. It is an inverted six with a single underhead camshaft and the exhaust ports on the left side, where they get cooling air before it reaches the intakes. As mentioned earlier, this is not the most efficient way to cool an engine, but perhaps it is a safer way to handle the peak heat loads that occur in aerobatic competition. The Avia has fuel injection, and so there is no more Zlin cough.

The use of air-cooled inlines was never as extensive, proportionately, in Italy as in other European countries. The first and most popular Italian inlines were the Colombos, made first by Colombo in 1929 and built by Alfa Romeo after 1931. Alfa may have bought out the original firm to secure the services of Ing. Colombo; he was a race car engine designer of the highest eminence. The first Colombo was an upright four of 347 cu in. (5.7 L) and 100 hp, a somewhat low specific output for an engine designed by a man who was used to racing automobile engines. A six, of 520 cu in. (8.52 L) and 130 hp, was added in 1930. Breda used Colombos in biplane trainers for the military and for a rakish-looking two-seater sport plane that looked rather like the Curtiss A-8 Shrike, CANT made them optional in several models, and Savoia-Marchetti put one in what was probably the most attractive light amphibian ever built, its two-place, cantilever-winged S-80.

Italian airframe builders were not nationalistic in their choices of engines. Not only did the use of air-cooled inlines decline in Italy after about 1934, but when one was used it was quite apt to be of foreign origin. Czech Walters were imported, and the British Gipsy was much used, notably by Caproni. Fiat may have encountered an actual preference for imported engines when it introduced its 135-hp A 60 in 1932, because it dropped it the next year. However, Isotta-Fraschini was more persistent.

Isotta introduced two air-cooled inlines in 1928 and continued to offer them as late as 1937 despite negligible demand. Its model 80 was an upright six with exposed valve gear, interesting chiefly in that it gave 33 percent more power—120 hp as against 90—when geared than it did in direct-drive form. The other Isotta, the Asso Caccia, was more unusual in that it was the only post-1918 air-cooled V-12 to be built in upright form. It either had very inefficient cooling or was intended for use in very slow aircraft, because it came equipped with a large air scoop that protruded above the top line of the cylinders like that on an RAF4. Every other air-cooled twelve built after 1918 was adequately cooled by the amount of air that could be admitted through the space between the vee of the cylinders, although the Ranger installation in the SOC-3 does come to mind as a case where this expectation may have been optimistic.

Perhaps one may be allowed to suspect that enthusiasm had as much to do with the Italian engine picture as did cold calculations of market potential. How else are the activities of a firm like the CNA (the ambitiously named Compagnia Nazionale Aeronautica) to be accounted for? CNA was headed by a prince and managed by a count. It built airframes and engines in small numbers, ran a flying school and a charter service, and probably raced Topolini on weekends. CNA built the highest-revving and most geared-down aviation engines ever made anywhere. One was a 12-cylinder X-type—four banks of 3 cylinders each—that developed 240 hp at 4900 rpm, and the other was a four that gave 90 hp at 4800. Both employed dual overhead cams, which makes eight camshafts for the twelve. This firm also built a trimotored airplane using three of its 90-hp fours.

The Inverted Inline Today

As was the case with the smaller radials, the development of air-cooled inlines in the free world ended with the Second World War. The inlines that could contribute to the war effort by powering trainers and secondary military craft—Rangers, Gipsy Majors, Arguses, Menascos, and Hitachis—stayed in production throughout most of the war.

In the United States, radials had replaced inlines in trainers by 1943. Apparently only the V-770 was produced beyond that year. My own opinion is that the only reason for producing any inlines at all in the United States during the war was that, at first, anything usable and in production had to be utilized. As soon as enough Kinners were available to replace the Menascos in Ryans and enough Continentals were available to let PT-23s take the place of the Ranger-powered PT-19s, both these inlines were taken out of production. While the facts are hard to pin down, apparently the radials required fewer work hours both to make and to maintain, in exchange for a slight performance gain that meant nothing for primary trainers.

After the war, defeat, poverty, and restrictions prevented the revival of the Hirth, the Argus, and the Hitachi. By the time private aviation revived in Germany, in Japan, and, to some extent, in France and in Italy, the American opposed four had developed into general aviation's dominant powerplant. Isotta made an abortive attempt to build a V-16, and some production of Regniers and Renaults as SNECMAs took place in France, but only in Britain and Czechoslovakia was there any extensive production of the old pre-war inline designs. Britain, in particular, turned out a substantial number of inline-powered Austers (anglicized and developed Taylorcrafts) before private flying there succumbed to "fair shares for all" and retreated to its present position as the coterie sport of the Tiger Club.

So these likable engines have just about died out. They were a little more expensive to build and maintain than were the opposed types, but this may not have been the chief cause of their decline. I suspect that they were done in by the tricycle gear. This works best with a small-diameter propeller, such as is called for by the high-revving flat engines, while the constant-speed propeller does away with much of the thrust advantage of a slower-running engine. As long as private flying included a good proportion of tandem-seat tail draggers with wooden propellers, the air-cooled inline was a viable choice; but those days are gone, and so is the type of engine that helped make so many of those old-timers such pretty airplanes.

Data for Engines

Maker's name	Date	Cylinders	Configuration	Horsepower	Revolutions per minute	Bore and stroke, in. (mm)	Displacement, cu in. (L)	Weight, lb (kg)	Remarks
RAF 1	1	8	V, A	90	1800(?)	3.9 (100) × 5.5 (140)	537 (8.8)	NA	
RAF 4	14	12	V, A	140	1800	3.9 (100) × 5.5 (140)	806 (13.2)	637 (290)	Usually described as 150 hp
RAF 4D	16	12	V, A	240	2000	4.1 (105) × 5.5 (140)	886 (14.5)	695 (316)	
ADC Airsix	19	6	I, A	300	1950	5.4 (139) × 7.5 (190)	1024 (16.8)	620 (282)	
Cirrus Mark I	25	4	I, A	65	2000	4.1 (105) × 5.1 (130)	274 (4.5)	268 (122)	
Cirrus Mark II	26	4	I, A	84	2000	4.3 (110) × 5.1 (130)	302 (4.9)	280 (127)	
Cirrus Hermes	28	4	Iv, I, A	115	2100	4.5 (114) × 5.5 (140)	347 (5.7)	300 (136)	Wildly varying output figures in different sources
Cirrus Major	35	4	Iv, I, A	135	2350	4.7 (120) × 5.5 (140)	383 (6.3)	310 (141)	
Cirrus Minor	35	4	Iv, I, A	90	2600	3.7 (95) × 5 (127)	220 (3.6)	205 (93)	
Cirrus Midget	38	4	Iv, I, A	55	2600	NA	NA	155 (70)	
Cirrus New Major	37	4	Iv, I, A	155	2450	4.7 (120) × 5.5 (140)	383 (6.3)	338 (153)	War apparently prevented production
de Havilland Gipsy I	27	4	I, A	100	2100	4.5 (114) × 5 (128)	318 (5.2)	285 (130)	Also called "Major Series III"
de Havilland Gipsy II	30	4	I, A	120	2200	4.5 (114) × 5.5 (140)	347 (5.7)	298 (135)	
de Havilland Gipsy Major	32	4	Iv, I, A	130	2350	4.6 (118) × 5.5 (140)	374 (6.1)	305 (139)	Gipsy III was same engine inverted
de Havilland Gipsy Six	32	6	Iv, I, A	203	2350	4.6 (118) × 5.5 (140)	561 (9.2)	486 (221)	Gipsy Queen similar
de Havilland Gipsy Twelve	37	12	Iv, V, A	525	2600	4.6 (118) × 5.5 (140)	1121 (18.4)	1058 (480)	
de Havilland Gipsy Minor	38	4	Iv, I, A	90	2600	4 (102) × 4.5 (115)	222 (3.6)	216 (98)	

Engine		Cyl.	Config.		rpm	Bore × stroke in (mm)			Remarks
Napier Rapier	30	16	H, A	305	3500	3.5 (89) × 3.5 (89)	539 (8.8)	720 (327)	
Napier E 97 Javelin	32	6	Iv, I, A	170	2300	4.5 (114) × 5.3 (133)	501 (12.7)	410 (186)	
Napier Dagger	34	24	H, A	890	4000	3.8 (97) × 3.75 (95)	1027 (16.8)	1390 (632)	
Renault 4 Pb	28	4	I, A	90	2000	4.5 (115) × 5.5 (140)	355 (5.8)	286 (130)	
Renault 4 Pci	32	4	Iv, I, A	120	2300	4.5 (115) × 5.5 (140)	355 (5.8)	308 (140)	
Renault 4 Pdi Bengali	33	4	Iv, I, A	145	2350	4.7 (120) × 5.5 (140)	386 (6.3)	386 (175)	
Renault 6 Pdi Bengali Six	34	6	Iv, I, A	180	2200	4.7 (120) × 5.5 (140)	587 (9.6)	451 (205)	
Renault Special	34	6	Iv, I, A	325	3250	4.3 (110) × 5.5 (140)	485 (8)	485 (220)	Gave 370 hp for 1935 races
Renault 12R. 01	38	12	Iv, V, A	450	2500	4.7 (120) × 5.5 (140)	1157	940	
Regnier R. 4	30	4	Iv, I, A	120	2100	4.7 (118) × 5.5 (140)	375 (6.1)	302 (137)	130 hp at 2350
Regnier R. 6	33	6	Iv, I, A	210	2500	4.5 (114) × 5.1 (130)	484 (8)	460 (209)	Coupe Deutsch racing engine; 6B03 same engine with Roots supercharger; 250 hp
Regnier 4 J	36	4	Iv, I	60	2300	3.5 (90) × 4.3 (110)	170 (2.8)	165 (75)	
Regnier 4 L	36	4	Iv, I, A	90	2150	4.7 (120) × 5.5 (140)	384 (6.3)	297 (135)	1947 version gave 145 hp at 2350 rpm
Regnier Regnier 12C-01	39	12	Iv, V, A	450	3200	NA	915 (15)	NA	War prevented production
Renault 4P 03	39	4	Iv, I, A	100	NA	NA	NA	NA	Some post-war production under SNECMA name
SNECMA-Regnier 4L-04	48	4	Iv, I, A	147	2340	4.7 (120) × 5.5 (140)	384 (6.3)	341 (155)	Post-war; design said to be simplified for easier manufacture
Chaise AV2	30	4	Iv, V, A	100	2000	4.3 (120) × 5.5 (140)	386 (6.3)	336 (153)	
Chaise 4 E	32	4	Iv, V, A	40	2500	3.4 (85) × 3.5 (88)	122 (2.0)	186 (85)	

(cont.)

Data for Engines (continued)

Maker's name	Date	Cylinders	Configuration	Horsepower	Revolutions per minute	Bore and stroke, in. (mm)	Displacement, cu in. (L)	Weight, lb (kg)	Remarks
Michel AM 14 type I	28	4	I, A	90	1850	4.5 (115) × 5.9 (150)	381 (6.2)	304 (138)	
Michel AM 14 type II	28	4	I, A	120	1850	4.9 (125) × 5.9 (150)	449 (7.4)	356 (162)	
Train	35	2	I, A	20	2300	3.2 (80) × 3.9 (100)	61 (1.0)	77 (35)	4- and 6-cylinder models had same cylinders
Allison V-1410	27	12	Iv, V, A	410	1800	4.6 (117) × 7 (178)	1410 (23)	1010 (459)	
Curtiss V-1460	30	12	Iv, V, A	525	2300	4.9 (124) × 6.5 (165)	1460 (24)	925 (420)	Experimental; only a few made
Dayton Bear	30	4	I, A	100	1500	4.5 (114) × 7 (178)	445 (7.3)	375 (170)	Converted from Hall-Scott with 5 × 7 Liberty cylinders
Dayton Aero Four	30	4	I, A	50	2500	3.9 (98) × 4.3 (108)	201 (3.3)	173 (79)	Available as kit for building from Model A Ford components
Fairchild 6-370	29	6	Iv, I, A	122	2100	3.9 (100) × 5.1 (130)	374 (6.1)	325 (147)	
Ranger 6-390	30	6	Iv, I, A	120	2150	4 (102) × 5.1 (130)	386 (6.3)	345 (157)	
Ranger 6-410	39	6	Iv, I, A	165	2450	4 (102) × 5.4 (138)	411 (6.7)	350 (159)	Stroke calculated from displacement; published dimension wrong
Ranger 6-440	41	6	Iv, I, A	175	2450	4 (102) × 5.5 (140)	440 (7.2)	350 (159)	
Ranger V-770	32	12	Iv, V, A	250	2150	4 (102) × 5.1 (105)	772 (12.6)	565 (257)	Introductory hp
Chevrolair D-4	29	4	Iv, I, A	120	2200	4.6 (117) × 5.3 (133)	352 (5.8)	345 (157)	D-6 had same cylinders, 180 hp
Phillips V-8	40	8	Iv, V, A	280	NA	4.6 (117) × 5.3 (133)	664 (10.9)	NA	Displacement as announced; does not check with bore and stroke
Michigan Rover R-236	2	4	Iv, I, A	55	1900	3.9 (98) × 5 (127)	236 (3.9)	210 (95)	
Michigan Rover R-267	30	4	Iv, I, A	75	1975	4.1 (105) × 5 (127)	267 (4.4)	232 (105)	

Heath B-4	29	4	Iv, A	30	3000	2.8 (70) × 3.5 (89)	83 (1.4)	119 (54)	Others also made
Tank V-502	31	8	V, A	115	1675	4 (102) × 5 (127)	502 (8.3)	NA	Henderson motorcycle engine conversions
Tank V-470	31	8	V, A	115	1650	3.9 (98) × 5 (127)	471 (7.7)	398 (181)	Made as late as 1937
Skymotor Model 70	38	4	Iv, I, A	60	2050	3.9 (98) × 4.3 (108)	201 (3.3)	186 (85)	
Menasco 4-A	29	4	Iv, I, A	90	1925	4.5 (114) × 5.1 (130)	326 (5.3)	265 (120)	
Menasco B-4	31	4	Iv, I, A	95	2000	4.5 (114) × 5.1 (130)	326 (5.3)	295 (134)	
Menasco C-4	31	4	Iv, I, A	125	2175	4.8 (121) × 5.1 (130)	363 (5.9)	296 (135)	Hp was 150 at 2260 rpm for C-4-S model
Menasco B-6	31	6	Iv, I, A	160	1976	4.5 (114) × 5.1 (130)	486 (6)	398 (181)	
Menasco 6CS-4	35	6	Iv, I, A	260	2300	4.8 (121) × 5.1 (130)	545 (8.9)	550 (250)	
Argus As. 8	28	4	Iv, I, A	110	2100	4.7 (120) × 5.5 (140)	386 (6.3)	248 (113)	
Argus As. 10	31	8	Iv, V, A	220	2000	4.7 (120) × 5.5 (140)	771 (12.6)	423 (192)	
Argus As. 16	33	6	Iv, I, A	230	2400	4.7 (120) × 5.1 (130)	538 (8.8)	352 (160)	
Argus As.410	39	12	Iv, I, A	450	3250	4.1 (105) × 4.5 (115)	728 (11.9)	660 (300)	
Hirth H-M 60	31	4	Iv, I, A	70	2150	3.9 (100) × 4.3 (110)	211 (3.5)	181 (82)	H-M 60R had 4 (102) bore, 80 hp
Hirth H-M 4	40	4	Iv, I, A	90	2400	4.1 (105) × 4.5 (115)	242 (4.0)	238 (108)	
Hirth H-M 150	32	8	Iv, V, A	170	2250	4 (102) × 4.3 (110)	400 (6.6)	330 (150)	
Hirth H-M 6	40	6	Iv, I, A	135	2400	4.1 (105) × 4.5 (115)	364 (6)	330 (150)	
Hirth H-M 8	33	8	Iv, V, A	245	3000	4.1 (105) × 4.5 (115)	485 (7.9)	539 (245)	First geared Hirth; met Coupe Deutsch requirements
Hirth H-M 12B	38	12	Iv, V, A	450	3100	4.1 (105) × 4.5 (115)	727 (11.9)	606 (275)	

(cont.)

Data for Engines (continued)

Maker's name	Date	Cylinders	Configuration	Horsepower	Revolutions per minute	Bore and stroke, in. (mm)	Displacement, cu in. (L)	Weight, lb (kg)	Remarks
Zlin Toma 4	37	4	Iv, I, A	106	2520	4.1 (105) × 4.4 (115)	244 (4)	209 (95)	
Walter Junior	32	4	Iv, I, A	105	2000	4.5 (115) × 5.5 (140)	355 (5.8)	292 (133)	
Walter Junior Major	33	4	Iv, I, A	120	2100	4.7 (118) × 5.5 (140)	375 (6.1)	312 (142)	Became Major 4 in 1934
Walter Minor	35	4	Iv, I, A	85	2260	4.1 (105) × 4.5 (115)	242 (4)	211 (96)	
Walter Major Six	34	6	Iv, I, A	200	2200	4.7 (118) × 5.5 (140)	563 (9.2)	385 (175)	
Walter Mikron	35	4	Iv, I, A	60	2600	3.5 (88) × 3.8 (96)	142 (2.3)	135 (61)	
Walter Minor Twelve	38	12	Iv, V, A	370	2750	4.1 (105) × 4.5 (115)	727 (11.9)	715 (325)	
Walter Saggita	37	12	Iv, V, A	500	2500	4.7 (118) × 5.5 (140)	1125 (18.4)	819 (372)	
Avia (Polish)	36	4	Iv, I, A	64	2450	3.7 (95) × 4.1 (105)	181 (3.0)	180 (82)	
Avia (Czech post-war) M-337	70	6	Iv, I, A	210	2750	4.1 (105) × 4.5 (115)	364 (6.0)	326 (148)	
Colombo S-53	28	4	I, A	94	1900	4.5 (115) × 5.5 (140)	374 (6.1)	226 (103)	
Colombo S-63	30	6	I, A	145	2050	4.5 (115) × 5.5 (140)	521 (8.5)	344 (156)	
Fiat A.60	31	4	Iv, I, A	140	2200	4.7 (120) × 5.7 (145)	400 (6.6)	302 (137)	
Isotta-Fraschini Asso Caccia	28	12	V, A	480	2250	4.9 (125) × 5.5 (140)	1257 (20.6)	838 (381)	
Isotta-Fraschini model 80 T	28	6	I, A	90	1650	3.9 (100) × 5.5 (140)	403 (6.6)	NA	
CNA C-X	32	12	X, A	240	4900	NA	NA	370 (168)	1725 rpm at propeller
CNA C-4	34	4	I, A	90	4800	3.2 (80) × 3.5 (90)	110 (1.8)	230 (105)	

V—V-type; A—air-cooled; Iv—inverted; I—inline; H—H-type; X—X-type; NA—not available. Open figures—U.S. Customary; figures in parentheses—metric. Weights are dry.

TEN

The opposed engine.

The Rationale of the Opposed Engine

Every piston engine in production today, outside of the Soviet bloc, is of the opposed type. It has displaced all others wherever buyers and designers are free to choose. Its merits lie not in any single area of outstanding superiority but in having an optimum combination of qualities for today's conditions and airframe designs.

The replacement of a radial by a flat engine in the Bellanca demonstrates this combination. In its pre-war incarnation as the model 14, the Cruisair was a two- to three-place airplane powered by a 95-hp Ken-Royce radial. By 1948 it had become a four-placer with a 150-hp Franklin opposed six up front. Eventually it went over 200 hp with a flat engine and acquired a tricycle gear. Visibility over the nose of the original model was barely acceptable; with a 150-hp radial it would have been impossible. A 150-hp radial would have had a diameter over the cowling of about 44 in. (1120 mm), more than the fuselage width and high enough to demand extra fuselage depth in order to let the pilot see past the engine. To begin with, then, both visibility and streamlining are better on small planes when opposed engines are used.

In addition, there is usually no weight penalty involved in the use of an opposed engine instead of a radial. A radial would be lighter for the same power if it had the same number of cylinders, but it doesn't; in the interest of smoothness, it needs more. A major source of vibration that plagues other types of engines is largely eliminated on opposed engines because their pistons come together and move apart in pairs, and this makes it practical to use fewer but larger cylinders. This, as we have noted earlier, makes for a lighter engine and allows the flat engine to be at least competitive with the small radial with regard to weight. Modern vibration-absorbing mounts help, of course. Because of both of these factors, we can now make flat fours in sizes that would have called for 7 cylinders in a radial.

The rpm of opposed engines tend to be higher than those of radials. This is a disadvantage from the standpoint of propulsive efficiency, but this is minimized by the availability of constant-speed propellers in the lower power ranges. In addition, the higher revs call for smaller propeller diameters, which go well with tricycle gear.

It can be summed up this way: Can you imagine a Bonanza with a radial engine?

An inverted inline would work better than a radial would in a modern small plane, and in Europe a few four-place "retractables" were designed for inlines. But, as a rule, visibility is still not as good, and the weight and cost are up. The inline does make for better streamlining in a narrow-bodied plane, but just about every post-war design has had side-by-side seating.

The 450- to 600-hp range saw the last stand of the radial. For 8- to 10-place planes like the Beechcraft DS-19 and the Percival Prince, only radials were suitable, and production of the Wasp Junior and the Alvis Leonideses continued into the sixties. But the turbines kept getting smaller and the flat engines kept growing, and, by 1970, the triumph of the opposed engine was complete.

Childhood and Adolescence of the Opposed Engine

It took a long time to grow up. Duthiel-Chalmers built opposed engines as long ago as 1905, but it was not until just before World War II than anyone made one big enough to power anything larger than a 1300-lb (591-kg) two-seater.

The first opposed engine built in respectable quantities was the Lawrance C-2 of 1917, intended not for real airplanes but for "Penguin" trainers. This engine was enough to set back 20 years any cause with which it was associated. It set at naught the basic advantage of the opposed engine by using a single-throw crank. With this Mickey Mouse arrangement, both pistons move in the same direction at the same time while, duly obedient to Newton's first law, the rest of the engine shakes violently in the opposite direction. With 28 hp available at 2000 rpm, it had a useful output for a light single-seater, but it would literally shake a 500-lb (227-kg) airplane to pieces. There was no possibility of using vibration-absorbing mounts because it had no mounts. It had to be attached to the airframe by means of metal straps clamped firmly around the cylinder barrels! However, they were cheap—$75 to $125 as war surplus—and so

one Orville Hickman devised the heroic solution of making a new crankshaft and bending the connecting rods to suit. Bending the rods shortened them, and so Hickman's conversion included cutting the rods and welding the parts together a little further apart. If you have ever wondered why "fearless" is sometimes equated with "stupid," consider the happy pilot flying his 1929 homebuilt over rough country behind a single-ignition wonder like this (Fig. 10-1).

The opposed engine got off to a better start in England. Due in large part to encouragement afforded by newspaper prizes and by government assistance to flying clubs, two rather good small twins came on the market during the early twenties. The first of these, around 1920,[1] was the ABC Gnat, made by the same firm that had built the disastrous Wasp and Dragonfly radials during the war. The Gnat used all-steel cylinders similar in design to those on the firm's radials, but they worked much better on the smaller engine. The Gnat gave, it was said, 45 hp at 1800 rpm from 139 cu in. (2.28 L), figures that, if honest, would have been considered excellent a decade later. In 1924 the Gnat was followed by the Scorpion Mark I of 24 hp and, a year or so later, by the Mark II of 34 hp. ABC survived long enough to bring out an engine that held considerable significance for the future, although this fact was not recognized at the time. This was its Hornet, apparently the world's first flat four. It was described as a "double Scorpion," though it had a longer stroke than either mark of the twin, and it gave 75 hp at 2175 rpm from 243 cu in. (4 L). It was, alas, ahead of its time. In the face of strong competition from the inline fours beginning to come from ADC (see Chap. 9), ABC left the aircraft engine business after 1926.

A little later, in 1923, the Bristol Cherub came out. It gave 25 hp at 2500 rpm, weighed 90 lb (41 kg), and displaced only 67 cu in. (1.1 L); a geared version weighed 107 lb (49 kg). In 1925 the Cherub III (the II was the geared version of the

---

[1]Designed earlier; produced post-war.

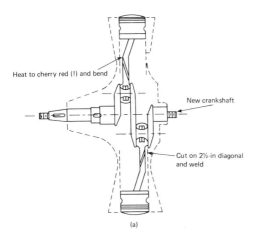

FIG. 10-1(a) The Hickman conversion of the Lawrance C-2 to a two-throw engine.

earlier model) replaced it, with 75 cu in. (1.23 L) and 36 hp at 3200 rpm available. The Cherubs had a peculiar valve-actuating system, somewhat like elongated Z-type rocker arms but set parallel with the cylinders. They were described, accurately enough, as "rocking shafts." Their inner levers were moved by the cams, and the outer levers moved the valves, which were set parallel to the crankshaft. While this odd system was never used on another engine, it worked very well, and the tiny Bristols were much respected engines for ultra-light single-seaters. Some were exported to the United States, where they won numerous races on such airplanes as the Heath Baby Bullet and the Powell P-H Racer.

America's first peacetime opposed twin came out in the same year as the Cherub. This was the Morehouse M-42. Giving 20 hp at 3000 rpm from 42.4 cu in. (69 cc), it was really too small to be useful. In 1925, Morehouse was taken over by Wright, and the next model, the WM-80, was called a Wright-Morehouse. As before, the model number was derived from the displacement (1310 cc) and the new engine had the more practical output of 28 hp at 2500 rpm. In contrast to the unnecessary refinement of aluminum cylinders with steel liners that helped make the Cherub an expensive beastie,

the Morehouse engines used cast-iron cylinders. Despite this, they weighed a bit less than the corresponding Bristols. Unfortunately, there was no place for such an engine in the OX-5-dominated world of U.S. private flying in the mid-twenties, and the Wright-Morehouse stayed in production for only a year or two.

In Germany the Treaty of Versailles created a rather more favorable environment for the minimum airplane, and there an engine much like the Morehouse achieved some degree of success. It came, somewhat surprisingly, from the big-engine house of Mercedes. Like the Morehouse, it gave 20 hp, but it betrayed its ancestry by weighing more than twice the American engine's 51 lb (23 kg). It went into the first Klemm, and, as the Klemm became Germany's premier ultralight, a respectable number were produced; the F 7502, as it was called, stayed in the catalog for at least 10 years after its 1922 introduction.

There was one other German engine of the same type, the Haacke. It was made in 28-hp and 30/35-hp forms and distinguished itself by being

FIG. 10-1(b) Lawrance C-2. Note mounting straps around cylinder barrels, just inside the cooling fins. On an aircraft the mounting saddles would be at the back, not below as here. (Courtesy of the author, taken at Glenn Curtiss Museum, Hammondsport, New York.)

no less than 43 in. (1090 mm) wide. Perhaps surprisingly to those not familiar with this firm's early products, it was used in one of Willy Messerschmitt's first power planes. Unlike the Mercedes, the Haacke had a short production life.

I think that, by now, the reason for the slow acceptance of the flat engine is beginning to emerge. It is simply that most people thought of it in terms of a 2-cylinder engine for minimum aircraft. The flat engine could not come into widespread use until either light aircraft achieved wider acceptance or the engines were made larger. The first of these happened (at least to some degree) early in Germany, never in Britain and France, and only with the Depression in America.

The first lightplanes to gain wide acceptance in the United States were the Aeronca Collegians, the single-seat C-1, and the two-seat C-2 and C-3. These triangular-fuselage planes probably took their inspiration from an Oregon-built homebuilt called the Harlequin Longster. To power this well-designed little plane, "Les" Long built a 30-hp, 90-lb (41-kg) opposed twin using Harley-Davidson 74 cylinder assemblies with a crankcase and crankshaft of his own design. It was called the Harlequin and was probably the best motorcycle-engine conversion ever built for aviation use, but its limitations—both technical and commercial—were obvious, and so Aeronca designed its own engine. Called the E-107, for its displacement (1.75 L), it was an L-head opposed twin giving 30 hp at 2500 rpm. Its specific consumption was, like many L-head aircraft engines, atrocious: 0.7 lb/(hp)(hr) [0.318 kg/(hp)(hr)]. It was one of the few engines ever built whose design was integrated with that of a specific aircraft having a deep, finned wet sump whose lines blended with the C-1's fuselage shape. The intake passed through this sump, keeping the mixture warm and the oil cool.

The E-107's output was inadequate for the contemplated two-place Collegian, so Aeronca's next move was to use a pair of Warner Scarab cylinder assemblies on the E-107's base. Displacement went up to 113 cu in. (1.85 L), leading to the designation of E-113, and power rose to a useful 40 hp. Installed on the Aeronca case, the airflow was at 90° to that of the Scarab installation, but they nevertheless cooled adequately. Whether be-

cause this peculiarity bothered someone or because of cost, the next, and definitive, Aeronca was the E-113A with Aeronca-designed cylinders.[2] The E-113A was produced in Britain by James A. Prestwich, the noted motorcycle-engine builder, as the Aeronca-J.A.P. (Fig. 10-2).

As long as Aeronca employed the E-113A in its C series planes, all was well. They were stable and cute, and their whole ambience was such that nobody expected any real performance from them. Leo Kimball, designer of the Beetle radial, was once asked to check out a C-3 whose owner felt that it was performing sluggishly. He flew a circuit, landed, and felt the engine: only 1 cylinder was warm. The owner had been flying with two people and only 1 cylinder! Presumably used to "real airplanes," the owner hadn't had any feeling that anything was too radically wrong.

But when the model K came out in 1939 in

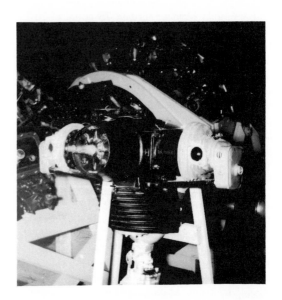

FIG. 10-2 Aeronca E-113A twin. America's first horizontally opposed engine to gain real popularity and, as such, the direct ancestor of every lightplane engine in production today. *(Courtesy of the author, taken at Glenn Curtiss Museum, Hammondsport, New York.)*

[2]Aeronca designed but did not make these engines. Manufacturing was done by Govro-Nelson, a noted Detroit machine shop.

response to competition from the Cub, Aeronca's 2-cylinder engine didn't provide good enough performance to sustain sales. The Scout, as the K-type was called, was a fine-looking little airplane and outperformed the C-type, but the trouble was that it looked like a real airplane and people expected it to act like one—a demand that they hadn't made on the C-type. The ultimate solution was a totally new airplane with 65 hp, but Aeronca's interim answer was to offer the Scout with two optional engines in place of the E-113A.

Best known of these was the Continental A-40, an engine whose importance cannot be overstated. It was the first popular flat four, and, as such, it is the direct ancestor of every piston engine in production today in the free world. The A-40 was the engine that made the classic Cub possible. Yet its distinctive design features were never used on another production engine. It was an L-head motor with cast-iron cylinders, one-piece heads spanning 2 cylinders, single ignition, and only two main bearings. Introduced in 1931, it gave 37 hp at 2500 rpm from 115 cu in. (1.9 L); the rare, later, twin-ignition model was rated for 40 hp. As might be expected from the low-efficiency L-head design, the specific consumption was a dismal 0.7 lb (0.318 kg), but this still gave the low total consumption of 2.8 gal/hr (10.6 L/hr) at cruise.

The single ignition was quite reliable, largely because a good spark plug location avoided fouling tendencies. The main bearings tended to wear bell-mouthed from the flexing of the unsupported crank, but this did not lead to trouble if the recommended TBO of 200 hr was adhered to. It was easy to start, smooth, reliable, and—more of a virtue during the early thirties than at any time before or since—cheap. A Scout with A-40 power cost less than the same plane with the 2-cylinder E-113A (Fig. 10-3).

Nothing did as much to keep private flying alive during the Depression as the combination of the Taylor—later Piper—Cub and the A-40 engine. It must be said, though, that the performance that the little Continental gave the sturdy (and concomitantly, heavy) Cub was poor. I can remember sitting over Rockaway Airport in Queens, New York, in a Cub on a windy day and having to study the terrain carefully to be sure that I was moving at all. It was boredom with flying in a J-2, all I could

afford, that made me abandon my power flying lessons at the age of 19.

The next significant flat four probably owed its conception to the limitations of the A-40. In 1935 the H. H. Franklin Co. of Syracuse, New York, an old builder of air-cooled automobiles, went broke, and a group of its engineers bought the factory at the bankruptcy sale. They reconstituted the firm as Air Cooled Motors Development Co. and put it into the aircraft engine business. Its first product, retaining the Franklin name, was the 4AC150. This was a considerably more sophisticated engine than the A-40, though I'm not sure that it was really a better one. It had steel-lined Y-alloy[3] cylinders made in one piece with the heads, three main bearings, and hydraulic lifters for its enclosed overhead valves. I have heard that it was sold as a bolt-in replacement for the A-40, but I cannot attest to this. At a nice 2150 rpm, it developed 50 hp from 150 cu in. (2.46 L) and weighed 156 lb (70.9 kg). A slower-turning 40-hp version was also listed, and dual ignition soon became available as an option.

The Franklin did not improve the Cub's performance as much as was hoped, though a fair number of Cubs were sold with Franklin power. However, in the lighter Aeronca it came into its own. A "KF" could match the performance of a 65-hp Cub. Unfortunately for Franklin, Continental came out in the same year—1938—with the A-50, the engine that the Cub had needed all along. But the Franklin company made a lot of 150-cu-in. engines anyway, for a most peculiar nonflying application.

The White Truck Company of Cleveland, recently freed by the court from its near-fatal association with Studebaker, had designed the first of the now well-known stand-up delivery trucks. It was called the White Horse and used an automotive version of the 4AC150 in one of the damnedest drive arrangements ever seen. White bolted the transmission rigidly to the differential and mounted

---

[3]Y-alloy was a British-orginated aluminum that had the same coefficient of thermal expansion as steel, most useful in cases where the differential expansion rates of steel and ordinary aluminum would cause problems when things heated up.

FIG. 10-3 Continental A-40. As the power plant of the first Cubs, this little engine did more to keep private flying alive during the Depression than any other engine. In terms of number of components, the A-40 was probably the simplest four-stroke aviation engine ever built. *(Courtesy of Teledyne Continental Motors Corporation.)*

the front of the engine on a big rubber ball, allowing the whole megillah to pivot as the axle moved up and down on the springs. Somewhere in this mass of misplaced ingenuity was a blower to keep the engine cool. Until people found out that a little air-cooled engine was not the ideal replacement for a milk deliverer's horse, White sold a lot of these things. Thus Franklin got to build a lot of 4AC150s, and by the time the White Horse passed from the scene, the firm was doing well with larger engines.

Lycoming introduced the third of the flat fours in the same year (1938). It was called the O-145 for its displacement and gave 50 hp at 2300 rpm. Its construction was somewhat of a tour de force, Lycoming having succeeded in casting all 4 cylinders in one piece with the iron crankcase. It used bolt-on aluminum heads, and the overhead valves sloped downward, making for one of the first examples of a wedge-shaped combustion

chamber. Like the Franklin, it had a simple-looking and somewhat shallow sump, but, while the Franklin's oil tended to get hot during a long climb, the Lycoming had no such problems. In fact, it was so trouble-free that it was quickly developed into a 65-hp, and later a 75-hp, version. Somehow, though, the Lycoming was never as popular as the Continental or the Franklin, and its cast-block construction was not continued on the firm's very successful later and larger models (Fig. 10-4).

Continental also introduced a new flat four in 1938, its A-50. At 171 cu in. (2.8 L) it was larger than its competitors and, because of this, could deliver its power at a low 1900 rpm. At 171 lb (78 kg) it was only slightly heavier than its competitors. There were to be no more shortcuts for Continental; the A-50 was a real airplane engine throughout. Its cylinders were built like a radial's with steel forgings screwed into aluminum heads; the crankcase was a split aluminum casting; and dual ignition was standard from the beginning. It cost more, but by 1938 things were sufficiently back to normal for performance to again outweigh cost. It employed a unique form of semi-dry-sump lubrication, having a baglike steel sump below the crankcase into which the oil drained without the need for a scavenger pump (Fig. 10-5).

Employed in a Cub or a Taylorcraft, the A-50 provided a real gain in performance and became for a short time the preferred powerplant for these popular lightplanes. However, Continental was not resting on its laurels, and within a year the all-time favorite,[4] the A-65, came out. It was really the same engine except perhaps for a different camshaft and carburetor. In the -3 version it was theoretically available with single ignition, but I never saw one and doubt that anyone else ever did. The A-65 became standard in the J-3, the

definitive Cub, and in the Champion and Chief with which Aeronca replaced the somewhat delicate Scout.[5] The A-65 delivered its power at 2300 rpm; the less popular A-75 was really the same engine cleared to run at up to 2650. The A-65 stayed in production until around 1960, and thousands of them survive on the grass strips where people still fly for fun.

I know of only two other fours from this era, and neither was successful. Harry C. Stutz, of Bearcat and HCS car fame, built a prototype in Chicago shortly before his death. I cannot find any data on it, but I remember it as having some sort of complex head design that resulted in an excessively wide engine. On the West Coast, Menasco saw the handwriting on the wall and tried to diversify into the opposed engine game. However, instead of building an engine based on cylinders from one of its inlines, Menasco tried to build a

---

[4]With apologies to Continental's O-200, which fits a different era of aviation with a different spirit.

[5]The Scout was strong enough in the air but suffered from "hangar rash" under careless ground handling.

FIG. 10-4 Lycoming's entry in the flat-four sweepstakes, the O-145. Overhead valves put it ahead of the A-40, but it used even more cast iron than the Continental in its construction. *(Courtesy of Lycoming Division, AVCO.)*

FIG. 10-5 Continental A-65, powerplant of the classic J-3 Cub, the Aeronca Champ, the Chief, and many others. Note the unique external sump. (*Courtesy of Teledyne Continental Motors Corporation.*)

simpler engine and wound up reinventing the Continental A-40 as a 50-hp engine with the same displacement as the A-40s. It had some sales, but what lightplane fliers were waiting for was a real little airplane engine, not a refined version of a power-mower motor.

Franklin and Lycoming enlarged their fours before Continental did. By the outbreak of the war Franklin was building its 4AC171 (as before, the designation was based on the displacement) in 65- to 90-hp versions with direct drive and in a geared version giving 113 hp at 3500 rpm. They were smooth-running, good-looking engines whose crankcases were a trifle weak and whose oil temperature had to be watched when climbing on a hot day. Franklin's best customer was probably Stinson, who built the first opposed-engine lightplane to break the two-place barrier. A Stinson 105, Franklin-powered, was the first lightplane I ever flew in that had a finished and trimmed interior like an automobile, complete with ashtray.

Franklin had also moved into a 6-cylinder series of 298 cu in. (4.9 L) and 130 hp with direct drive; geared, 175 hp could be had. The opposed

engine was growing up (Fig. 10-6). Actually, there had been some large opposed engines earlier, but none of them was commercially successful. Ashmusen, in Providence, Rhode Island, had built a flat eight of 65 hp and a twelve of 105, in 1908: to help cool the cylinders, intake air was passed over them on its way to the carburetor. In 1929, Charles B. Kirkham built a quite large and rather good-looking opposed six; I do not know how powerful it was, but my guess is that it developed well over 150 hp. There is one, dated 1934, in the Glenn H. Curtiss Museum in Hammondsport, New York, that is built to run vertical rather than flat. But these were premature engines: however good they may have been, there were no airplanes suitable for them, and they did not offer enough advantages to lead anyone to design airplanes around their then-unconventional shapes.

These Franklins used the same type of construction as the original 4AC150, but when Lycoming went to larger engines they abandoned the *en bloc* construction of the O-145. The first of the new Lycomings, the O-235, was a 100-hp four whose construction was very much like the Continental's, but without the bladderlike oil reservoir. This construction became standard with the succeeding Lycomings: the O-290 four, the 150-hp O-350, and the 175-hp O-435 sixes. Without the downward-protruding sump, these engines could be somewhat more neatly cowled than could the Continental, and Continental dropped the deep adjunct sump on its later engines.

Thus, by the time civilian production in America was halted by the war, the opposed engine had finally moved out of the minimum-airplane class. The airplanes that used tbe newly enlarged engines fell into two categories: developments of the familiar strut-braced high-wing types and precursors of the post-war low-wing retractables. The leader in the first group was the Stinson mentioned above. Originally a marginal three-placer with a 75-hp Franklin, it got 90 hp in 1940. Revised after the war as the 108, it became a genuine four-place airplane behind a 150 Franklin six and eventually went to 165 hp.

Piper did something similar with the J5C Super Cruiser, a 75-hp, three-place outgrowth of the Cub. It went to 90 hp, then to 100, became a four-seater as the Family Cruiser, and eventually got to as much as 150 hp. In the course of these changes, Lycomings gradually came to displace Continentals in much of the Piper line. This was a natural development in at least one sense, because Piper's home base in Lock Haven, Pennsylvania, is only 27 mi (43.5 km) down the road from Williamsport, where the Lycomings are made.

First of the low-wing retractables was the Culver Cadet, an Al Mooney design that is remembered with mixed feelings. It was a hot little rock with an elliptical wing of only 27-ft (8.2-m) span that claimed 140 mph (225 km/hr) on 65 hp. The Cadet was exciting to land, not only because of the high wing loading but also because the gear operation was a little uncertain. I had a ride in one that involved several sharp pullouts of at least two g's to get the gear unstuck. During the war, Culver built a single-seat version called the PQ-14 as a radio-controlled target drone; this was probably the first low-wing single-engine aircraft to have a retractable tricycle gear, unless one counts a special model of the radial-engine Spartan Executive that never went into production.

Other early low-wing retractables exploring the line of development that has resulted in the present dominance of the flat engine were built by Piper and Aeronca. By the time the war stopped production, the signs were clear. Post-war America was going to be opposed-engine country.

FIG. 10-6   The 335-cu in. (5.5-L) Franklin six. The squarish shape of the cooling fins was long a Franklin characteristic. This engine was also built in a vertical-shaft, blower-cooled version for helicopters. *(Courtesy of NASM-Smithsonian.)*

The Opposed Engine in Europe

Meanwhile, considerable development was taking place on the Continent. Argus, in 1932, built an engine in the class of the Continental A-40. It was called the As 16 and, thanks to the use of overhead valves, was more efficient than the Continental; its specific consumption was 0.572 lb (0.26 kg), and it used a three-throw crank. This was an idea that had been tried earlier in Europe in an experimental Anzani, with one difference that is worth illustrating (Fig. 10-7).

In 1935 the French Mengin (sometimes given as Menguin, but this is incorrect) firm began to build a line of opposed twins that were a little too small to become popular. Its type B gave 28 hp,

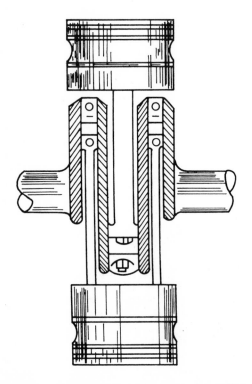

FIG. 10-7 How the experimental Anzani twin worked—lunatic logic applied to the design of a smooth two-cylinder engine. As the cylinders are in line instead of staggered as would be the case with a two-throw crank, maximum smoothness is achieved. Now, why couldn't Orville Hickman have thought of this?

the type C 38, and the type GHM 45. The contemporary Belgian Saroleas were similar in concept; the Epervier gave 25 hp, the Albatross 30, and the Vautor 32—as misnamed a trio of little engines as can be found anywhere. A few Mengins and Saroleas must have been built, because their names are occasionally encountered in connection with *Pous de Ciel* and post-war homebuilts, but I can find no record of any production aircraft using them. The opposed engine was truly successful in only one European country: Czechoslovakia.

Three Czech firms, Jawa, Walter, and Praga, made opposed engines. The Jawa, built by an important motorcycle firm, was geared off the camshaft at half engine speed and gave 35 hp at 4000 rpm. Most aircraft engines built by motorcycle builders were derived from their bike engines, but this doesn't seem to have been the case with Jawa. Its aircraft engine was a four-stroke, whereas its motorcycle had two-stroke power. The Jawa was listed as late as 1938 but was little used. The Walter Atom, like the Jawa a midget twin, was a direct-drive engine giving 28 hp at 3000 rpm and did not gain anything like the acceptance accorded Walter's inlines and radials.

The Pragas were both larger and more successful. The first was a 36-hp, 2150-rpm twin, the 1936 B2. As was the case with the Aeronca (though less aesthetically effective in the Praga's case), this engine and a particular airplane were designed for each other. The aircraft, the Praga E-114, was an extraordinarily interesting one. It was a cantilever high-wing monoplane having the unique feature[6] of side-by-side seating in front of the wing. Access was provided by having a 2-ft (0.6-m) section of the leading edge swing up with the canopy attached, and the occupant's heads rested against the main spar. The great advantage was the sailplanelike visibility, but special means had to be used to prevent this forward seating from resulting in a nose-heavy aircraft. These included building the wing with all taper at the trailing edge and a straight leading edge (making for a forward-swept lifting line), using a reflexed airfoil section, and keeping the engine weight as close in as possible.

[6]Until the coming of the Andreasson BA 7/Bolkow-Malmo Junior of 1970.

The B2 helped in this because, being a twin, it was inherently short and the occupant's feet could be tucked under the rear of the engine. The whole combination worked out extremely well. The little Praga had the best performance ever for a plane in its power class and held numerous records. It was made under license in England as the Hillson-Praga, as was the engine (by Jowett of Bradford, the old Yorkshire builder of opposed-engine cars and trucks).

The other Praga was a four of 174 cu in. (2.85 L) called the D and the DR. Its output was listed by its makers as 79 hp at 2610 rpm, but there is some uncertainty about this, as a later derivative of the E-114 listed its powerplant as a "65-hp Praga." 79 hp is a little high for the displacement but 65 is low for the rpm given. It was a little different in construction from most fours, having its heads made in pairs, each spanning 2 cylinders. Continental did this on the A-40, but the Continental's "flatheads" were much simpler than the overhead-valve type of the Praga. In addition, it used four cams—one for each cylinder.

Praga's major post-war effort, the Doris, took a different tack. It was quite a big six, displacing 437 cu in. (7.2 L) and delivering 220 hp at 3000 rpm. The Doris had overhead cams with rocker arm covers partly incorporated in the cam covers, giving a very characteristic branched appearance. It was built through the late fifties, largely to power spray planes, but was dropped after 1961. In addition to the Praga, there were two post-war Czech engines, officially identified only by number, that may have been Pragas under the skin. One, called the M 110H, was a 4-cylinder engine built to run with its shaft in a vertical position for use in light helicopters. A 210-cu in. (3.3-L) engine that delivered 120 hp at 3100 rpm, it could only have been used in very light copters. Its companion, the M 108H, was a six that gave 270 hp and was said to be in production both in horizontal- and vertical-shaft models. Both of these engines appeared in 1961 and, as far as we in the West have been informed, disappeared by 1965. Walter is also said to have made a new flat four for a year or so, but I have no data on it.

Poland also produced some ephemeral opposed engines after the war. PZL began to develop a 65-hp engine called the WN-1 very shortly after the war, but it was dropped in the early fifties. Then, in 1961, the Narkiewicz design group picked it up and made it into the NP-1, giving 68 hp at 2600 rpms from 176 cu in. (2.88 L). This was followed by the 6-cylinder 220-hp, 421-cu-in. (6.9-L) WN-6. Both these engines were dropped by 1969. The WN-6 was apparently used in the first models of the strange Wilga plane but was replaced by a radial in the production version.

In France, the Potez firm developed what might have become a widely used series but which apparently succumbed to the *defi Americain*. Potez's first flat engine was the 4E of 1960, a 209-cu-in. (3.42-L) four of 105 hp at 2750 rpm. It saw considerable use in some of the many and confusing Jodel derivatives that everybody seemed to be making in France for a while. In 1961 it added the 6E, a 155-hp fuel-injected six of 314 cu in. (5.13 L), but the whole thing came to an end in 1964 when Potez became a part of Lycoming's parent company, AVCO.

There was another post-war French effort that could equally well be listed under radials or inlines. The industrialist Emil Mathis returned to France from the United States and built a 100-hp inverted inline four, a 210-hp inverted V-8, a 175-hp 7-cylinder radial, a 500-hp 14-cylinder radial, a 40-hp opposed twin, and a 75-hp opposed four, all using the same cylinder. The basic cylinder had a bore of 3.78 in. (96 mm) and a stroke of 3.94 in. (100 mm), resulting in a displacement of 723 cc (44.17 cu in.). Thus a range of engines from 88 to 618 cu in. (1.45 to 5.07 L) was possible, but this very interesting idea came to nothing.

The very eminent Italian motorcycle manufacturer, Agusta, also had a shot at the opposed aircraft engine game. Agusta's first engine was the 2-cylinder G.A. 40, an engine of somewhat unusual appearance. The pushrods were in front instead of below (like most such engines) or on top (like Potez's). This made it look like a radial with only 2 cylinders (Fig. 10-8). A remarkably efficient design, it gave 42 hp at an acceptable 2700 rpm from a displacement of 91 cu in. (1490 cc) and weighed only 104 lb (47 kg). Had this design been amenable to scaling up, Agusta might have had a real future in aviation, but the cam design made it difficult to

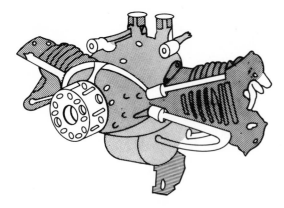

FIG. 10-8   The Agusta G.A. 40 twin, probably the last opposed two-cylinder engine to be developed. A good engine but too late; by 1953 there was little market for such a low-powered engine. The splayed-out pushrods are conspicuous.

double it into a four.[7] Thus Agusta's four, when it came out, had few components in common with the G.A. 40 and was a less impressive engine. The G.A. 70, as it was called, gave 86 hp at the low weight of 150 lb (68 kg), but it had the fatal flaw of being a direct-drive motor that had to turn 3500 rpm to achieve its rated power. The company's material said that buyers could attach their own reduction gears if they wished! The Agustas were off the market by 1969.

Two other, largely nonindustrial, countries have tried their hands at making opposed aircraft engines. The Hindustan Aircraft Company of Bangalore, India, after reinventing the Aeronca Chief as the Hindustan Pushpak, went on to similarly reinvent the Lycoming flat four to power it. Called the P.E. 90 H, it ran on 72-octane gas and used largely materials of Indian origin. Its designers do not seem to have been handicapped by these restrictions, because they got 90 hp at 2600 rpm from 192 cu in. (3.15 L) and a weight of 182 lb (82.7 kg). It was announced in 1960, but a curtailment of the production plans for the Pushpak

resulted in a decision not to put the P.E. 90 H into production. Too bad.

The other surprising small engine was the Spanish ENMA Flecha. It had the resources of the old Elizalde group behind it, but Spanish manufacture was handicapped, much as Indian was, by the absence of some of the advanced alloys available elsewhere. Announced in 1965, this 90-hp four had been dropped by 1969.

A Pragmatic Sanction in Britain

The British have dropped their former support for flying clubs and through oppressive regulation have created difficulties for those enterprising enough to pay for their own flying. Nevertheless some general aviation continues to exist in the land of "fair shares for all"; much of it involves company-owned planes, but there is still something of a market for private aircraft and for engines to power them. One by one during the post-war years, most of the former British engine manufacturers have been brought under the Rolls-Royce, Ltd., umbrella. As the suitability of the existing inverted inline light-plane engines gradually diminished, Rolls was faced with the problem of finding more up-to-date replacements for them. The demand would hardly justify designing a new line of opposed engines which, in any event, would be hard put to compete with the low prices that resulted from the large-scale production of the American engines. Rolls, therefore, took the sensible step of taking out a license to build Continental engines in Britain.

Rolls has built a wide variety of Continental models, ranging up from the smallest, the C-90 (some of which were exported to India to power Pushpaks!), through the ubiquitous O-200 to the turbocharged GTSIO-520 C of 340 hp. Equally interestingly, there has been some mutually beneficial exchange of expertise. Continental credits Rolls with a major input into the development of its latest series of engines. Rolls still has to fight for what business it gets; several British planes continue to list Lycoming power as standard. The magic of a name does not sweep everything before it (Fig. 10-9).

The Rolls-Continental engines need not have

[7]Nothing that would have fazed Hirth, of course; Hirth would have simply added more cams and pushrods at the back!

been the only opposed engines in Britain. After the war, Sir Roy Fedden oversaw the design of a flat six that applied the lessons he had learned in 10 years of building sleeve-valve radials at Bristol. The Fedden was an impressive and professional-looking engine, and I, for one, will always regret that it was never put into production. In view of the demonstrated virtues of the single-sleeve-valve[8] type, such an engine would have constituted a real challenge to the existing types. Unfortunately, Fedden had got himself spread out too thin, money was tight, and Fedden died before things could be straightened out.

There has been one more opposed engine made, if that is the word, in England, the Rollason-Ardem. The fact is that it, and the Belgian Limbach, the German Pollman, the American Revmaster,

and their cousins, are all Volkswagens. The Volkswagen firm has been willing to supply engines to convertors but has never endorsed their use in aircraft. This being the case, it seems more appropriate to examine them in the next chapter under Automobile Conversions.

Post-War America: The Triumph of Continental and Lycoming

At the end of World War II, the American aircraft industry was in an excellent position to gain a lasting worldwide dominance of the lightplane powerplant business. We had suffered no war damage, our designs were up to date, and our production facilities were efficient. U.S. engine builders were ready to go, and the customers had the money. Equally significant for the long run was the fact that the large American market would justify mass-production tooling on a scale hardly economical elsewhere. Finally, an era of freer international trade was dawning.

Two of the three significant American build-

---

[8]In fairness, it must be said that not everyone agrees on these virtues. Rob Meyer of NASM feels that the Bristol sleeve-valve engines were "unnecessary"; Prof. Rogowski of MIT believes that their friction is inherently excessive; and so on. Nevertheless I stand on the record. Other things equal, sleeve valves give about 1 extra hp per 15 cu in. (250 cc) of displacement.

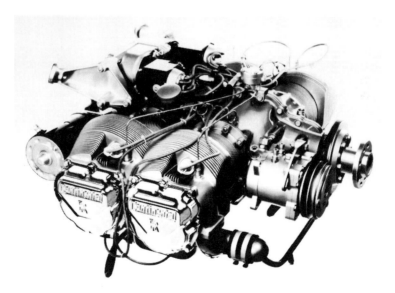

FIG. 10-9   Rolls-Royce-built Continental I0-368. This engine delivers as much as 220 hp in turbosupercharged form and is rated at 1800 hours TBO. The fuel-injection arrangements show up very clearly in this picture. (*Courtesy of Teledyne Continental Motors Corporation.*)

ers of horizontally opposed engines exploited these advantages to the fullest. As a result, Continental and Lycoming do in fact dominate the general aviation piston engine market in the free world today. The third major builder of the type, Franklin, was less fortunate.

Franklin's troubles largely grew out of furnishing the powerplant for the Republic Seabee amphibian. The biggest optimists in America in 1946 were the aviation people, none more so than those at Republic. Republic thought that its four-place Seabee was a plane that could be produced and sold by the thousand. Unfortunately, there was an engine-supply problem. The Seabee used a 215-hp Franklin six, and Franklin couldn't provide engines as fast as Republic could build Seabees unless it neglected all its other customers, which it wasn't about to do. So Republic bought into Franklin, and all was well until the Seabee bubble burst. Now Franklin, quickly cut loose by Republic, had no customers. Some builders of Franklin-engine planes had gone out of business in the same slump that had hit Republic, and others had recertified their airplanes for other engines when they couldn't get Franklins. Franklin could only struggle along, barely surviving on a little helicopter business, while Continental and Lycoming went from strength to strength.

Lycoming entered the post-war period with a line comprising three fours, of 65, 100, and 125 hp, and a six of up to 240 hp. As early as 1947 the helicopter market was being catered to with a vertical-shaft, blower-cooled version of the six, rated at 225 hp. In 1948 the first flat eight was added, the GSO-580 of 578 cu in. (9.5 L) and 320 hp; the opposed engine was beginning to move in on radial country. The venerable iron-cylinder O-145 was dropped in 1953; by that time, what 65- and 75-hp engine business was left was going to Continental. Also in 1953, Lycoming added an all-time winner—the big (150-hp) 4-cylinder O-320. In 1955, the O-435 six was bored out from 4 ⅞ in. (114 mm) to 5 ⅛ in. (130 mm) to make the O-480, with up to 280 hp available at takeoff. Two even larger fours, the 170-hp O-340 and the 180-hp O-360, were added in 1957. A 541-cu-in. (8.87-L) six, using the same cylinders as the O-360 and giving 250 hp, joined the line in 1959. This extension of the big-cylinder scheme to the sixes resulted in an engine that was only 40 cu in. (656 cc) smaller than the earlier eight, still in the line but not selling especially well.

The largest of all the flat Lycomings so far has been the O-720, an eight based on O-360 cylinders and introduced in 1964. This, it turned out, was getting too close to turbine territory to have a large sale. More popular has been the TIGO 541, a developed version of the O-540, with up to 450 hp available. This engine, incidentally, illustrates the most recent tendency in the design of the larger flat engines: it is turbosupercharged for better high-altitude capabilities (Fig. 10-10).

A dynamic development history like Lycoming's implies that business has been good, but this, in turn, raises the old question of the chicken and the egg. It is quike likely that business has been good because the firm (and, to at least as great a degree, its principal competitor) has vigorously pursued every promising line of development. It is anybody's guess whether Franklin could have made a comeback if it could have developed a more varied line.

The Franklin line contracted while those of Lycoming and Continental expanded. In 1946 the Syracuse company offered 176- and 199-cu-in. (2.88- and 3.26-L) fours and 298- and 405-cu-in. (4.88- and 6.64-L) sixes. Then, in 1947, these were replaced by a 225-cu-in. (3.7-L) four and sixes of 335 and 500 cu in. (5.5 and 8.2 L). The small six, in a vertical-shaft model, was quite popular for helicopters; the big one, with a cooling blower and an extension shaft for pusher installations, was the Seabee's engine. The big six was replaced by one of 425 cu in. (6.96 L), apparently built only as a helicopter engine, in 1953. The four died out by 1956, and the model 425 in 1959.

In 1961 Franklin became a part of Aero Industries. This was a group put together by an entrepreneur named Alexandre Berger, involving Jacobs, the corporate shell of Waco, and a few others. The four was revived as the 4A225 of 125 hp, and a twin, called the 2A110, using the same cylinders, was introduced. The following year the displacements of both small engines underwent a slight enlargement. The twin, now called the 2A120 and giving 60 hp, was adopted as the standard powerplant for a revived version of the old Aeronca Champion; its maximum cruising speed was a

FIG. 10-10 Lycoming TIGO (for *T*urbocharged, *I*njected, *G*eared, *O*pposed) 541-E1A flat six, 425 hp from 541 cu in. (8.87 L). This picture is worth a thousand words in explaining the demise of the 9-cylinder radial in its last stronghold, the 400-plus hp class. *(Courtesy of Lycoming Division, AVCO.)*

sparkling 75 mph (120 km/hr). In 1967 the six also got big-bore cylinders and the four became available in kit form for homebuilders. The twin was dropped in 1969 and revived in 1970. But nothing helped. Aero Industries went definitively broke, and in 1977 the plant, designs, and rights were sold to Poland, where the full Franklin line is apparently being produced today by PZL. The moral is that when emissaries from big firms show up with stories of the advantages to be gained from joining a conglomerate, you should show them the door at once.

Continental's post-war history is hardly less of a success story than Lycoming's. Its post-war lineup was similar: three fours and three sixes. The fours were the A-65, the C-75, and C-85 [the C-75 and C-85 were the A-65 bored out to give 188-cu-in. (3.1-L) displacement]. The sixes were the C-115, the C-125 (which was the C-115 rated for 200 more rpm), and the C-140, all designated by their rated output. In 1947 the C-140 was dropped and the A-100, a small six using A-65 cylinders, was added. Also in 1947, Continental introduced its first big six, the 471-cu-in. (7.6-L) E-165 and E-185 pair. Continental's line in the early post-war years was generally characterized by lower-rated speeds and lower specific outputs than Lycoming's, and its market share was probably smaller.

In 1948 the C-140 was dropped and the C-90, of 201 cu in. (3.28 L), and a companion six, the C-145, were added. With the C-145, Conti-

nental engine speeds began to rise; its rated 145 hp was delivered at 2700 rpm. The A-100 also came to the end of its road in 1948. No more changes, except for some redesignating of existing models, took place until 1953, when the O-315, a big four originally developed for the military, joined the line. In the same year the 471-cu-in. six began to undergo a change of designation. The E-185 version of this engine was continued (up to 1959) under the same number, but the rest of them were renamed O-470, with different "dash numbers" giving the clue to the rated horsepower.

The change to O designations took until 1970 to be completed. The C-75 and the C-125 were dropped in 1953, but the C-85 was listed up to 1966, the A-65 until 1967, and the C-90 until 1970. Meanwhile, the GSO-526, a big six, had been introduced in 1955, and the following year the C-145 became the O-300A. Continental's big winner, the O-200A, came out in 1959. Developed from the old C-90, it was lightened and simplified a bit and had its rating raised to 100 hp at 2700 rpm. It was not outstanding in any way, to judge from its specifications, but it was reliable, long-lasting, and, most important, was the right engine at the right time (Fig. 10-11).

In 1959 the two-place field was suffering from neglect; every plane in the class was outdated. Cessna filled the gap with its 150, which could be described either as a modernized 140 or as a scaled-down 172. Without the O-200, there could

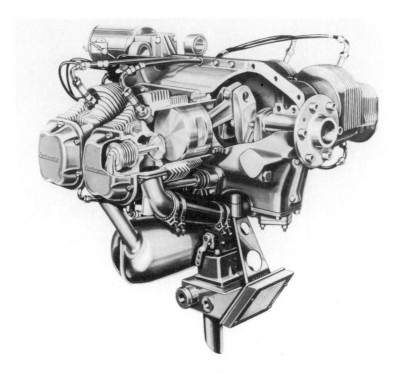

FIG. 10-11   The Continental 0-200, a candidate for the title of the best-selling lightplane engine ever built. *(Courtesy of Teledyne Continental Motors Corporation.)*

have been no Cessna 150; anything smaller couldn't have coped with the added drag of the nosewheel gear and the weight added by the flaps and the fancy interior, and anything bigger would have raised the cost too much. So many 150s were sold that, when O-200 production finally stopped in 1979, Continental claimed that over 100,000 had been manufactured. This figure, which would make the O-200 the most popular aircraft engine of all time, seems awfully high and may include remanufactured engines. It is probable that more Lycoming O-320s actually went into airplanes.

Continental, in 1971, became a part—perhaps the nucleus is a better term—of something called "Teledyne," but it did not lose its entrepreneurial spirit. The firm came out with the only really new series of "general aviation" engines in a long time, the Tiara line. These were relatively small, high-revving sixes that reverted to the old system of using beefed-up timing gears for reduction gearing. A cushioning connection took care of the torque-variation problem. The Tiara specifications were impressive. The 6-285 delivered 285 hp at 4000 engine rpm and weighed 382 lb (173 kg), and the Tiara 320 delivered 320 hp at the same weight. I do not have their fuel consumption specifications, but, as they had compression ratios of 9 to 1 and 9.5 to 1, respectively, to take advantage of 10- to 130-octane fuel, and as the low propeller speed gave high propulsive efficiency, they should have used considerably less gas than, say, an O-520 of the same power (Fig. 10-12).

This undertaking did not achieve the hoped-for success. Few airframe builders were willing to go to the expense of redesigning and recertificating their products, preferring to adopt a wait-and-see attitude as long as existing designs were selling well. The Tiara installations that were made tended to

FIG. 10-12 **This might have been the shape of the modern piston engine. An experimental Tiara eight, never put on the market. Note the high thrust line resulting from driving off the camshaft. Over-the-top cooling airflow and fuel injection were inherent parts of the design.** *(Courtesy of Teledyne Continental Motors Corporation.)*

justify this hesitation: there were recurring operational problems. After putting many millions of dollars into the Tiara program, Continental dropped the line in 1978.

In the meantime, Lycoming had been modernizing its line through a program of detail redesign, and Continental, deprived of the market advantage expected from the Tiara, was feeling the pinch of this vigorous competition. Thus, in 1979 the firm announced a new line of engines, developed with Rolls-Royce's cooperation. While the new Continentals break no new ground in terms of basic concepts, they do represent a very high level of engineering refinement. They are about 10 percent lighter than the engines they replace and are being certificated for as much as 2000 hr TBO from the start. The new Continentals come in two sizes, both based on the same cylinder design: a six of 520 cu in. (8.5 L) and a 346-cu-in. (5.7-L)

four called, for some reason, the 368 series. All have high compression ratios and use fuel injection; turbosupercharging is available on both the four and the six. While this was being written, the full power range had not been announced but up to 325 hp will be available (Fig. 10-13).

Thus, the opposed engine, unlike the other types we have examined, is very much a viable proposition today. It has a future. But to attempt to assess that future calls for extrapolation, which, as the designer of the de Havilland Comet said after he found out why Comets were coming apart in the air, is the fertile mother of error.

For example, the smaller engines are out of production as a result of steadily increasing power requirements for general aviation aircraft. This has come in part from demands for higher performance and in part because it has been cheaper to go to larger engines than to use refined structural and

FIG. 10-13(a)   Probably the most highly refined general aviation engine in the world today, the Continental TSIO-520 of up to 325 hp. Compare with Fig. 10-10: is the somewhat more finished appearance of the four the result of Rolls-Royce traditions and pride, or does Fig. 10-10 show an engine that has been specially finished up for exhibition use?

aerodynamic design in the airframe. Who will be so rash as to say that this trend will continue? If it reverses itself, we may see a trend to smaller engines. In addition, there are two opposing trends with regard to electronic equipment. On the one hand, the government, in its ongoing effort to foolproof the entire universe, keeps thinking up new kinds of gear that has to be carried. On the other, the makers of such equipment keep reducing its weight, bulk, and power requirements. Which will prevail, and what effect will the result have on aviation's power needs?

And all this applies only to general aviation's current mainstream. Will the current interest in foot-launched ultralights result in the reinventing of open-cockpit slow flying off grass strips, with perhaps a whole new set of relaxed requirements for off-airways aviation? If so, we should see some much smaller engines in the future.

This uncertainty applies to many other things besides engine size. The pursuit of fuel economy is very apt to lead to such features as stratified combustion, turbocompounding, and the Diesel. Despite the Tiara's failure, we should be seeing more and more high-speed geared engines. But whatever happens, it is long odds that the aero engine of the future will be of the opposed configuration.

FIG. 10-13(*b*)　Perfected engine guts. Parts on the left are from the new Continental series, those on the right from the previous model. Metal has been removed wherever it was not contributing strength and the cylinder cooling fins have been modified in accord with studies of the temperature zones in an operating cylinder. The net result has been a 10 percent weight reduction. *(Courtesy of Teledyne Continental Motors Corporation.)*

Data for Engines

Maker's name	Date	Cylinders	Configuration	Horsepower	Revolutions per minute	Bore and stroke, in. (mm)	Displacement, cu in. (L)	Weight, lb (kg)	Remarks
Lawrance A-3	16	2	O, A	28	1400	NA	NA	NA	Propeller from this engine is right for an 85-hp, 2600-rpm modern opposed four
ABC Gnat	13	2	O, A	45	1800	4.3 (110) × 4.7 (120)	139 (2.3)	115 (52)	Hp very high for size in 1913 if accurate, but ABC record suggests it is optimistic
ABC Scorpion Mark I	24	2	O, A	24	2500	3.6 (92) × 3.6 (92)	73 (1.2)	90 (41)	Mark II had 90 cu in., gave 40 hp at 2750 rpm
ABC Hornet	25	4	O, A	82	2175	4 (102) × 4.8 (122)	243 (4)	219 (100)	
Bristol Cherub I	24	2	O, A	34	4000	3.4 (85) × 3.8 (95)	67 (1.1)	81 (37)	Cherub II larger
Morehouse M-42	23	2	O, A	20	3000	3 (76) × 3 (76)	42 (0.69)	51 (23)	
Wright-Morehouse W-M 80	25	2	O, A	28	2500	3.8 (95) × 3.6 (92)	80 (1.3)	85 (39)	
Mercedes F 7502	19	2	O, A	20	3000	2.9 (75) × 3.9 (100)	54 (0.88)	106 (48)	Had 3 to 1 planetary reduction gear
Haacke HF M2	NA	2	O, A	30/35	NA	4.7 (120) × 5.5 (140)	193 (3.2)	132 (60)	A later Haacke gave 45 hp at 1400 rpm
Long Harlequin	32	2	O, A	30	2850	3.5 (89) × 5 (127)	96 (1.6)	90 (41)	Cylinders said to be from Harley 74 motorcycle; stroke may have been increased
Aeronca E-107	30	2	O, A	30	2500	4.1 (105) × 4 (102)	107 (1.8)	114 (52)	
Aeronca E-113	31	2	O, A	40	2520	4.2 (108) × 4 (102)	114 (1.9)	113 (51)	E-113A had same dimensions and performance
Continental A-40	31	4	O, A	37	2550	3.1 (79) × 3.8 (95)	115 (1.9)	144 (65)	The rare dual-ignition model gave 40 hp
Franklin 4AC150	38	4	O, A	50	2150	3.6 (92) × 3.6 (92)	150 (2.5)	163 (74)	
Lycoming O-145	38	4	O, A	50	2300	3.6 (92) × 3.5 (89)	145 (2.4)	152 (69)	

Engine	Year	Cyl	Type	hp	rpm	Bore × Stroke			Remarks
Continental A-50	38	4	O, A	50	1900	3.9 (98) × 3.6 (92)	171 (2.8)	170 (77)	A-65, 65 hp at 2300, and A-75, 75 hp at 2750, were same basic engine
Menasco M-50	38	4	O, A	50	2550	3.5 (89) × 3.8 (95)	144 (2.4)	156 (71)	
Franklin 4AC171	39	4	O, A	60	2350	3.9 (98) × 3.6 (92)	171 (2.8)	170 (77)	
Franklin 6AC298	40	6	O, A	130	2600	4.3 (106) × 3.5 (88)	298 (4.9)	295 (134)	
Lycoming O-235 -C	40	4	O, A	100	2600	4.4 (111) × 3.9 (98)	233 (3.9)	207 (94)	
Lycoming 0-290-A	41	4	O, A	125	2600	4.9 (124) × 3.9 (98)	289 (4.7)	244 (111)	
Lycoming 0-350	41	6	O, A	180	2700	4.4 (111) × 3.9 (98)	352 (5.8)	289 (131)	
Lycoming O-435-A	41	6	O, A	190	2550	4.9 (124) × 3.9 (98)	434 (7.1)	364 (165)	Geared O-435-A gave 240 hp, O-435-D gave 225 hp
Ashmusen	09	12	O, A	105	1800	3.8 (95) × 4.5 (114)	596 (9.8)	345 (157)	
Argus As 16	32	2	O, A	40	2500	3.2 (82) × 3.7 (95)	251 (4.1)	156 (71)	
Mengin type B	35	2	O, A	25	2280	3.7 (95) × 3.5 (90)	76 (1.2)	77 (35)	Also built as "Poinsard" make
Mengin type C	35	2	O, A	38	2500	4.1 (105) × 3.5 (90)	95 (1.6)	82 (37)	
Mengin type GHM	38	2	O, A	45	2400	4.1 (105) × 4.3 (115)	116 (1.9)	99 (45)	
Sarolea Epervier	34	2	O, A	27.5	3650	3.2 (81) × 3.5 (90)	56 (0.92)	109 (50)	
Sarolea Albatross	37	2	O, A	35	2950	3.5 (88) × 3.5 (90)	67 (1.1)	95 (43)	
Sarolea Vautor	35	2	O, A	32	2750	3.5 (88) × 3.5 (90)	67 (1.1)	NA	
Jawa	NA	2	O, A	35	4000	3.3 (84) × 3.5 (90)	61 (1)	86 (39)	
Walter Atom	38	2	O, A	28	3000	3.3 (85) × 3.8 (96)	66 (1.1)	88 (40)	
Prage B2	38	2	O, A	46	2510	4.1 (105) × 4.3 (110)	116 (1.9)	106 (48)	

(cont.)

Data for Engines (continued)

Maker's name	Date	Cyl-inders	Config-uration	Horse-power	Revolutions per minute	Bore and stroke, in. (mm)	Displace-ment, cu in. (L)	Weight, lb (kg)	Remarks
Praga D	37	4	O, A	79	2160	3.7 (95) × 3.9 (100)	174 (2.9)	142 (65)	DR was same engine but geared
Praga Doris	56	6	O, A	220	3000	4.5 (115) × 4.5 (115)	437 (7.2)	436 (198)	
M 110H	61	4	O, A	120	3100	NA	201 (3.4)	NA	
M 108H	61	6	O, A	270	NA	NA	NA	NA	
PZL WN-1	48	4	O, A	65	2500	3.9 (100) × 3.6 (92)	176 (2.9)	128 (58)	Weight seems too low
NP-1	61	4	O, A	68	2600	3.9 (98) × 3.5 (90)	166 (2.7)	NA	PZL became Narkiewicz
WN-6	61	6	O, A	220	3300	4.9 (125) × 3.7 (94)	421 (6.9)	346 (157)	Also called Narkiewicz
Potez 4E	55	4	O, A	105	2750	NA	209 (3.4)	NA	
Potez 6E	61	6	O, A	155	2700	NA	314 (5.1)	NA	Had fuel injection
Agusta G.A. 40	53	2	O, A	42	2700	3.9 (100) × 3.7 (95)	91 (1.5)	104 (47)	
Agusta G.A. 70	56	4	O, A	86	3500	3.9 (100) × 3.7 (95)	140 (2.3)	150 (68)	
Mathis	52	4	O, A	92	2650	4 (105) × 3.9 (100)	211 (3.5)	251 (114)	Bore slightly enlarged over that of Mathis radials and inlines
Hindustan P.E. 90H	60	4	O, A	90	2600	4 (102) × 3.8 (95)	192 (3.1)	182 (83)	
ENMA Flecha	65	4	O, A	93	2600	4.7 (120) × 5.5 (140)	211 (3.4)	323 (147)	
Fedden Flat Six	47	6	O, A	185	3400	4.3 (109) × 3.8 (95)	325 (5.3)	310 (141)	
Lycoming GSO-580-C	48	8	O, A	375	3300	4.1 (105) × 3.9 (98)	578 (10.4)	560 (254)	
Lycoming O-320	53	4	O, A	150	2700	5.1 (130) × 3.9 (98)	320 (5.2)	272 (123)	
Lycoming GO-480-A1A6	55	6	O, A	340	3400	5.1 (130) × 3.9 (98)	480 (7.8)	492 (223)	Direct-drive model less powerful

(cont.)

Engine		Cyl.	Type	Power	rpm	Bore × stroke	Weight		Remarks
Lycoming O-340	57	4	O, A	170	2700	5.1 (130) × 4.1 (105)	340 (5.6)	275 (125)	
Lycoming O-360	57	4	O, A	180	2700	5.1 (130) × 4.4 (111)	361 (5.9)	282 (128)	
Lycoming O-540	59	6	O, A	250	2575	5.1 (130) × 4.4 (111)	542 (8.9)	396 (180)	As TIGO-541, gives up to 450 hp
Lycoming O-720	64	8	O, A	520	2650	5.1 (130) × 4.4 (111)	722 (11.9)	597 (271)	
Franklin 4AC199	40	4	O, A	90	2500	4.3 (106) × 3.5 (88)	199 (3.3)	190 (86)	
Franklin 6AV405	46	6	O, A	245	NA	NA	NA	NA	No further data after first announcement; probably not produced
Franklin 4A4-B3	47	4	O, A	100	2550	4.5 (114) × 3.5 (89)	225 (3.7)	228 (104)	Became model 225 in 1948
Franklin 6A4-B3	47	6	O, A	150	2600	4.5 (114) × 3.5 (89)	335 (5.5)	365 (167)	Became model 335 in 1948
Franklin 6A8-215-B8F	47	6	O, A	215	2500	5 (127) × 4.3 (108)	500 (8.2)	485 (220)	Fan cooling and extension shaft for pusher installation; Seabee's engine
Franklin model 425	53	6	O, A	300	3275	4.8 (121) × 4 (102)	425 (6.8)	352 (160)	Vertical-shaft model for helicopters was most common
Franklin 2A120	64	2	O, A	60	3000	4.6 (114) × 3.5 (89)	118 (1.9)	130 (59)	Was 2A110, 4.5-in. (114-mm) bore, in 1963
Franklin 4A235	64	4	O, A	130	2800	4.6 (114) × 3.5 (89)	235 (3.9)	240 (109)	
Franklin 6V350	67	6	O, A	235	3200	4.6 (114) × 3.5 (89)	350 (5.7)	329 (149)	Helicopter engine only
Continental C-85	40	4	O, A	85	2575	4.1 (103) × 3.6 (92)	188 (3.3)	181 (82)	Same dimensions for C-75; basis of 190-cu.-in. racer class
Continental C-115	46	6	O, A	115	2350	4.1 (103) × 3.6 (92)	282 (4.6)	257 (117)	Same dimensions for C-125
Continental C-140	46	6	O, A	140	3000	4.1 (103) × 3.6 (92)	282 (4.6)	298 (135)	Geared version of C-115
Continental A-100	47	6	O, A	100	2350	3.9 (98) × 3.6 (92)	256 (4.2)	223 (101)	

Data for Engines (continued)

Maker's name	Date	Cyl-inders	Config-uration	Horse-power	Revolutions per minute	Bore and stroke, in. (mm)	Displace-ment, cu in. (L)	Weight, lb (kg)	Remarks
Continental E-165	47	6	O, A	165	2050	5 (127) × 4 (102)	471 (7.6)	335 (152)	Compare with specifications of Warner Super Scarab 165; E-185 had same dimensions
Continental C-90	48	4	O, A	95	2625	4.1 (103) × 3.9 (98)	201 (3.3)	188 (85)	0-200A had same dimensions, gave 100 hp at 2750 rpm
Continental C-145	48	6	O, A	145	2700	4.1 (103) × 3.9 (98)	301 (4.9)	258 (117)	
Continental O-315A	54	4	O, A	150	2600	5 (127) × 4 (102)	315 (5)	261 (119)	
Continental GSO-526	55	6	O, A	340	3100	5.1 (130) × 4.3 (108)	526 (8.6)	540 (245)	
Continental O-300A	57	6	O, A	145	2700	4.1 (103) × 3.9 (98)	301 (4.9)	277 (126)	
Continental Tiara 6-285	72	6	O, A	285	4000	4.9 (124) × 3.6 (92)	406 (6.7)	382 (174)	
Continental 520 series	79	6	O, A	325	2700	5.3 (133) × 4 (102)	520 (8.5)	380 (173)	Weight estimated
Continental 368 series	79	4	O, A	180	2700	5.3 (133) × 4 (102)	346 (5.7)	260 (118)	Weight estimated; output is for IO-368

O—horizontal opposed; A—air-cooled; NA—not available. Open figures—U.S. Customary; figures in parentheses—metric. Weights are dry.

chapter
ELEVEN

Lost causes, oddballs, and unconventional engines.

There has never been any scarcity of ingenious people who have thought that they have found a better way to power airplanes than a conventional engine, and a fair number of them have raised enough money to bring their ideas to trial. Some of these ideas have been tried a time or two, found to be dead ends, and dropped. Others have kept recurring as new people have come along who have thought they saw where earlier experimenters went wrong.

Barrel Engines

Consider the barrel engine, which is a powerplant with its cylinders parallel to the shaft instead of at right angles to it. Its attraction lies in the fact that its frontal area is at a minimum, since its cylinders form a sort of closely packed bundle. There seem to have been three problems with barrel engines, two of them technical and one psychological. The most basic technical problem is to find a good way to have pistons drive a shaft that lies parallel to them. This has usually involved either a swash plate or a barrel cam (Fig. 11-1). The second problem has been cooling, especially if the engine is air-cooled; there is a lot of closely packed machinery, most of which generates heat, in a barrel engine. The psychological problem is one that has doomed not only most barrel engines but many other radical types as well. Their inventors seem to be incapable of using known low-risk technology for all but the essential new element; they throw in untried valve designs, peculiar bearings, strange cooling systems, and whatnot. They are unable to resist the temptation to try to be heroes. In addition, they make outrageous claims for their brainchildren.

The Alfaro, a swash-plate type built by the Indian Motorcycle Company, of

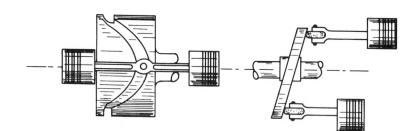

FIG. 11-1   The barrel cam and the swash plate.

Springfield, Massachusetts, in 1938,[1] exemplifies this. Not content to be a barrel engine, it was a two-stroke as well; it had funny valves involving rotating plates with holes in them; and it claimed the entirely incredible specific consumption of 0.435 lb (0.197 kg). The British government had traveled the same road earlier with the 1913 Statax, which, not content to be a barrel type, was at the same time a rotary. The Statax name stayed around long enough to be attached to an ambitious privately funded 1929 program involving sleeve-valve swash-plate engines, this time nonrotary. The U.S. Army Air Corps looked into so many barrel engines that, by the time it gave up on them in the early thirties, there was a stack of blueprints of unsuccessful engines that reached the rafters in a hangar at Wright Field.

None of this should be condemned as a waste of money. It should be remembered that for at least 60 years before Whittle, Ohain et al., people had been designing gas turbine engines that either melted or were hopelessly inefficient. Unless someone puts money into experiments, how will we ever know which wild-sounding ideas will turn out, with appropriate development work, to be worldbeaters? And who better than governments, considering the military benefits when the occasional breakthrough occurs?

Some of them, of course, might better have been drowned at birth. For instance, the 50-hp American Macomber was a swash-plate rotary on which the tilt of the swash plate could be varied to change the stroke of the pistons. This is done today on oil pumps and hydraulic motors to vary the displacement, but why anyone would want to do it on an airplane engine is obscure at best.

At least four barrel engines found alternatives to swash plates and barrel cams. The British Redrup and Fury engines used a "Z" crankshaft, set at an angle to the engine's axis. By having the pistons twist as they reciprocated, with the aid of a wristlike joint in the connecting rods, rotary motion was obtained (Fig. 11-2). The Gadoux and the Cleveland, on the other hand, were actually assemblies of more or less conventional engines, grouped in barrel form and driving a central shaft through bevel gears. Optimistically anticipating military use, the Gadoux had a hollow shaft to allow a gun to be fired through it.

What is interesting about the long list of unsuccessful barrel engines is that some excellent pumps have been made on this principle. Have a look some time at a Frigidaire automobile air conditioning compressor. Nicely engineered and simple as it is, adding cooling, exhaust, ignition, and valves for four-stroke operation would turn it into a mechanical nightmare. There may be echoes of this in the Wankel engine; it is derived from what is said to have been an effective pump, but it does

---

[1] Manufacture by Indian in 1938 was alone enough to guarantee failure. *De mortuis nil nisi bonum* and all that, but Indian was in those late days the farthest thing from a precision shop that could be found this side of the steam-locomotive factories.

FIG. 11-2 The British Wooler barrel engine of 1937, a swash-plate type with back-to-back pistons. In this type of swash-plate drive, ball-jointed connecting rods link the pistons with disks that are free to tilt but not to rotate. The disks, in turn, are mounted on the swash plate itself by means of ball bearings; the changing tilt of the disk causes the swash plate to rotate. Essentially this is the same as the so-called Z-shaft design. *(Photo. Science Museum, London.)*

not seem to have been nearly as successful as a powerplant.

The Caminez

One of the most publicized unconventional engines of its time was the 1925 Caminez, or Fairchild-Caminez. This was a 4-cylinder radial whose drive shaft carried a two-lobe cam instead of a crank (Fig. 11-3). As each cylinder fired, a roller on the lower end of the piston pushed the cam around for a quarter turn. As the cam turned, the rising lobes pushed up the pistons of two other cylinders, one on the exhaust stroke and the other on compression. Links pulled the opposite piston down on the intake stroke. It can be seen that the shaft revolved once for each two strokes of the pistons, giving the effect of a 2 to 1 reduction drive for the propeller and eliminating the need for cam gearing. The Caminez showed considerable promise, although—

like the Siemens rotary—it required a large four-bladed propeller. It passed government tests, but in Navy trials it vibrated excessively due to inherent dynamic imbalance and broke down due to inaccurate machining of some of its internal parts.

Unconventional Piston-to-Crank Connections

Some engines used conventional crankshafts but drove them in unconventional ways. Outstanding among these was the Gobron-Brillè. This engine had two opposed pistons in each cylinder, with the valves and spark plug in the middle. The lower one was connected to the crank in the usual way, but the upper one had a rod that ran to an outdoor crosshead (like a steam engine's) from which two additional connecting rods descended to two additional crank throws. Readers may think that

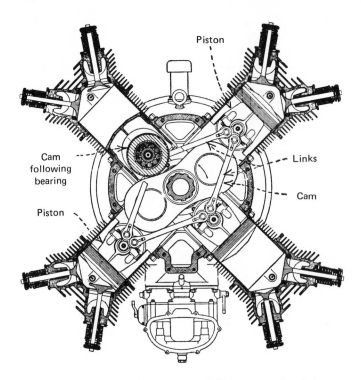

FIG. 11-3 The Caminez crankless engine. Unlike a normal radial, an even number of cylinders did not lead to an uneven firing order; 1,2,3,4 was inherent to the layout. *(Courtesy of NASM-Smithsonian.)*

there's something wrong either with this description or with their own understanding, but there isn't; the engines really were made this way. The marvelous thing about it is that it worked very well. Gobron-Brillè cars were almost unbeatable in *voiturette* racing for several years. The aircraft version was built around 1910 as a water-cooled 4-cylinder radial of very large diameter.

As an aircraft engine, the Gobron-Brillè did not fill any long-felt wants, but the idea was revived in 1929 in the United States. The Dawn was a 5-cylinder air-cooled radial on exactly the same principle. It used the crankcase of a 5-cylinder Velie, but it is not clear how the builders got the extra crank throws into the Velie crankcase. It should be noted that an air-cooled opposed-piston engine is almost certainly a loser because, without a well-finned cylinder head, most of the heat of combustion will go into the piston heads and melt them.

An equally odd opposed-piston engine was the little 18-hp Italian Beltrame of the late thirties. It had the distinction of being the only engine ever built to have 1 cylinder but two crankcases and crankshafts. A motorcycle chain tied the two shafts together. Being air-cooled, it probably didn't work too well, but it must have been the best-balanced single ever built.

Another way of tying the pistons to the cranks was represented in the design of the pre-war French Sega, one of the very few internal-combustion trunnion engines ever built. The trunnion engine is an old steam-engine scheme that has the whole cylinder rocking on a pair of biggish bearings instead of using wrist pins on the connecting rod. It didn't work terribly well on steam engines either. As if this wasn't enough, the Sega was a rotary as well. Yet even this was rational compared with the French Edelweiss rotary of

1913. The Edelweiss had an outer casing, to which the pistons were attached. They did not reciprocate: the cylinders did! It was the cylinders, of course, that were connected to the crankshaft (Fig. 11-4).

Double-Acting Engines

Internal-combustion engines are normally single-acting, meaning that their pistons are pushed down but not back, while steam engines are mostly double-acting, with the steam pushing the piston both back and forth. The double-acting idea has engendered some experiments, most interesting of which was the Deeble Duplex, built in California in 1937. It used a clever but complex arrangement of a piston within a piston, to do away with the need for a crosshead and to keep the height down. A Deeble was actually flown, in a Ryan S-T, but it never reached the market.

The fatal flaw in a double-acting internal-combustion engine is that it's not too healthy to

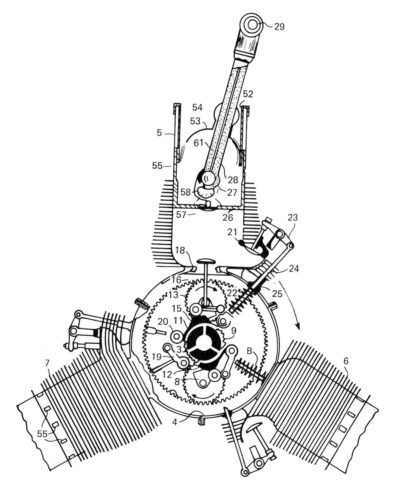

FIG. 11-4 An unidentifiable early "inside-out" rotary, believed to have been designed by the notorious "Mad Anthony" Mainbearing while smoking something illegal. *(Courtesy of A. O. Sherman.)*

have combustion heat on both sides of a piston. The oil in the crankcase of a single-acting engine cools the bottom of the usual "trunk" piston, but there is no comparable heat dissipation for the disk piston of a double-acting engine. Another difficulty is that the diameter of the piston rod is a deduction from the piston area and, hence, from the displacement of the lower half. This makes the downstroke more powerful than the upstroke and causes vibration. Some steam engines used a sort of dummy piston rod, called a "piston guide," in the upper half to bring things back into balance, but this introduces sealing problems where it goes through the cylinder head.

All these problems and more were (brilliantly) solved in the 1913 French Demont rotary. The Demont used hollow pistons, hollow piston rods, and hollow piston guides. The stroke was so short, and the diameter of the hollow rods so large, that wrist pins were located in the guides—eliminating the need for a crosshead. There was a disk-shaped baffle in the piston, and both the piston rod and the piston guide had open ends. When the engine ran, air passed through the whole assembly, cooling the piston from inside (Fig. 11-5). Whether this

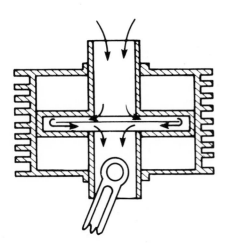

FIG. 11-5 Simplified cross section of cylinder and piston of the Demont rotary. Arrows show cooling airflow during the outward stroke of the piston; the flow would reverse during the inward stroke and would have a general outward bias from centrifugal force.

marvel of ingenuity was ever run is not ascertainable, but it was exhibited at the 1913 Paris Air Show.

Unclassifiable Inspirations

It has occurred to some inventors that a rotary could uncover valve ports as it rotates. First to try this, as far as I can determine, was the 7-cylinder Hermann, built in Philadelphia in 1925.[2] Another, the Mawen, first tried in France and continued in the United States after 1940, had the backing of some big names: Charles L. Lawrance of Whirlwind fame, the refugee French industrialist Mathis, and Axel Wenner-Gren, the Nazi-loving Swedish ball bearing king.

The Mawen had a ring containing manifolds and ports, surrounding the whole engine. The cylinders were open-topped and rode against this ring. As the engine revolved, the cylinders were exposed in turn to exhaust ports, intake ports, and a blank surface. Sealing was a problem, complicated in the case of the Mawen by differential thermal expansion—cool intake manifold, warm cylinders, and hot exhaust manifold. Since nothing was heard of the Mawen after 1942, it can be assumed that its problems were never solved.

The American Bakewell Wingfoot and Universal were radial equivalents of the Cleveland and Gadoux barrel types. Each was an eight that actually comprised four V-twins, each with its own crankshaft and all geared to a common central propeller shaft. Something similar was done in the sixties, a homebuilder making a "radial" comprising nine complete McCulloch chain saw engines.

One very interesting engine, the French Jalbert of 1929, pursued the attractive but elusive stratified-mixture concept. On top of each cylinder was a smaller cylinder, complete with piston, in which a too-rich-to-fire mixture was compressed. At the same time, plain air was compressed in the main cylinder. At the point of firing, a valve opened to allow the two to mix and fire. The total amount of air was greater than was needed for complete combustion, assuring very complete use of fuel,

[2]Not the same as the Herrmann, a 1937 barrel engine.

good economy, and low emissions. The best-known Jalbert was a Diesel, but I believe there was a gasoline version also. The Jalbert concept is one that might well be looked into again for automotive use, though part-throttle operation might be a problem.

Two-Stroke Engines

Two-stroke engines constitute the largest category of unconventional aircraft engines. As described in Chap. 2, once the four-stroke engine had attained adequate reliability, there remained only the appeal of low initial cost to justify a two-stroke. This has seldom been enough to sell significant numbers of any two-stroke. Nevertheless, several such engines stayed on the market for a number of years in Europe. This was possible because they were made by motorcycle-engine builders for whom they formed a minor but worthwhile sideline. Such engines tended to have many components in common with their makers' other products.

Hans Grade, the first German to fly, was one of these builders. In 1934 he returned to aviation with a 30-hp upright inline four. Another motorcycle-based engine was the Diecke, an 18-hp twin whose washing-machine aura was enhanced by the use of a flywheel magneto.[3] The Diecke claimed the specific consumption of 0.607 lb (0.276 kg), very good for a two-stroke engine. The Diecke had about the same specifications as the earlier (1926) Krober and may have been the same engine under a new name.

The English Scott was made by the same people who made one of the nicest motorcycles ever, the Flying Squirrel, and was named after the bike. It was a vertical twin giving a normal output of 16 hp at 2800 rpm and a maximum of 34 at 5200. The Scott, and the similar French Aubier et Dunne, were built to cater to the *Pou de Ciel* craze that swept France and Britain in 1934 – 1935. While the *Pou* was a death trap when equipped with heavy engines like the Cardan-Ford conversion, it

was safe enough with the 85-lb (39-kg) Scott or Aubier et Dunne (Fig. 11-6).

DKW[4] made both motorcycles, air-cooled, of course, and two-stroke water-cooled cars. Somewhat strangely, it was the car engine that it chose to offer in an aircraft conversion. It used the Schnurrle loop scavenging system that is only now coming into general use on American two-strokes. A 2-cylinder DKW that gave 20 hp and drove through a 2.5 to 1 reduction gear was used on the 1933 Erla single-seat lightplane; this was one of the very few two-stroke engines to be used on a production aircraft. There also seems to have been a 3-cylinder "Deke," giving 21 hp at 3000 rpm. These engines were essentially the same as those in the SAAB car between 1956 and 1971.

These were all three-port engines, in which the mixture enters the crankcase through a port in the cylinder wall that is uncovered when the piston reaches the top of its stroke. The French AVA, built for a few years after 1935 by yet another motorcycle-engine maker, got away from this system to

[4]You're not going to believe this. DKW originally stood for *Damp Kraft Wagen,* meaning "steam car." Then the firm built bicycle conversions and changed it to *Das Knaben Wunsch,* "The Boy's Desire" and, finally, to *Das Kleine Wunder* when it began to make cars.

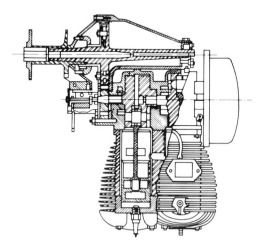

FIG. 11-6 A more practical engine for 17-hp airplanes than the Aubier et Dunne would be hard to find, but how practical could a 17-hp airplane be?

[3]References to washing machines may mystify those under 50. Before the spread of rural electrification, Thor and others sold a lot of gasoline-powered washing machines to farm users.

some extent. It was a flat four giving 28 hp at 2800 rpm and had a gear-driven rotary fuel-distribution valve located below the crankcase. It also had the peculiarity of single ignition with two magnetos, one for the right side and one for the left. Perhaps one was supposed to be able to make it to the nearest airport on 2 cylinders if one side conked out!

There were two German rotary-valve two-stroke engines. One, the 1932 inverted twin Von Festenberg-Pakisch, combined a rotary valve with the usual intake port in the cylinder wall. This was a configuration that reappeared in the United States in hot McCulloch Kart engines in the sixties. The other, the Schliha, was an opposed twin that had more design innovations than any engine ought to have to cope with. The cylinders were made of "Elektron" magnesium alloy sprayed with steel to provide a wearing surface. The bearing surfaces of the "Elektron" crankcase got the same treatment. The cylinder heads had little auxiliary cylinders built into them, and the pistons carried extensions, sort of little pistons, upgesticken from their tops. Der extensions in the auxiliary cylinders rode, also cutting down on side loads on the pistons und kompressing some of the kraftstoffe the cylinders to help charge. It will be noted that this design resulted in the only doughnut-shaped combustion chambers ever used in a production engine. The Schliha weighed 73 lb (33.2 kg), was 39.37 in. (1000 mm) wide, and, with the aid of two carburetors and two magnetos, gave a lusty 23 hp. It was available to intrepid builders for several years after its 1932 introduction.

Another version of the stepped-piston idea appeared in the English Caunter of the late twenties. This was a 5-cylinder radial. There is no possibility of having crankcase compression, necessary to a normal two-stroke, on a radial; as one piston moves in, others move out. The Caunter proposed to overcome this by having arc-shaped bottoms on the cylinders, sealing against matching pieces on the connecting rods. Had this improbable design worked, the small volume under each piston would have resulted in a desirably high base compression. Not content with this, the Caunter added stepped pistons to get what was described as "a degree of supercharge."

Results must have been disappointing, because a later Caunter design, the 1937 Shackleton model C, was a more conventional inverted four. This 60-hp engine used plain aluminum cylinders without liners and claimed fuel consumption of 3.5 gal/hr (14 L/hr, if these are Imperial gallons) at cruise. If "cruise" meant 45 hp, the specific consumption would have been under 0.5 lb (0.227 kg); no two-stroke engine ever attained such a figure without resorting to fuel injection.

One way to improve two-stroke engine efficiency is to use paired cylinders that share a common combustion chamber and have the intake ports in one cylinder and the exhaust in the other. The exhaust ports open first, allowing some blow-down of the pressure, then the incoming mixture pushes the exhaust out smoothly with a minimum of mixing. Puch (Allstate in the United States) used this system successfully for years. At least one aircraft engine used this idea, the 1929 Wichita Blue Streak. In appearance a 10-cylinder radial, it was actually five Puch-type paired-cylinder engines geared to a central shaft in the manner of the Universal and the Wingfoot.

This was not the only American two-stroke engine. In 1917, C. Harold Wills, then with Ford and later the builder of the noted Wills-Ste. Clair car, designed an inverted V-4. It was used to power an early missile, the Sperry Aerial Torpedo. While this may not be an airplane in the strictest sense, it is noteworthy that the air-cooled two-stroke in America ended up where it began—as a power-plant for pilotless aircraft. The last such U.S. engines were flat four and six McCullochs, made for radio-controlled target drones. The 72-hp McCulloch fours were used on Bensen gyrocopters, where their high speed (4000 rpm) went well with the small propellers needed for these craft. They were reliable in the hands of operators who understood them, but not everyone liked them (Fig. 11-7).

Most impressive of the U.S. two-stroke engines was the Hurricane radial. A supercharged, but not fuel-injected, 8-cylinder two-row engine, it came with a nice-looking streamlined crankcase and gave 150 hp. The specific consumption was claimed to be 0.55 lb (0.25 kg) at full power and 0.31 lb (0.14 kg) at cruise, impossibly low figures.

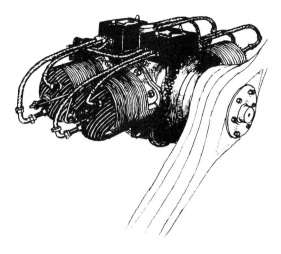

FIG. 11-7 McCulloch 4-cylinder two-stroke, 1962.
This engine had fuel injection to improve idling and
fuel consumption and oil metering to avoid the need
to mix oil and gas, but the result of these refinements
was a price that was too high for success—80 hp at
4100 rpm, 115 lb. (*Courtesy of the author.*)

Two American two-strokes went into actual
airplanes. In the late twenties, the same Brownback
who was involved with the Anzanis offered two
small two-strokes. The Tiger Kitten was an inverted
twin of 25 hp, and the Tiger Jr. was the same
engine doubled to make a 50-hp flat four. My
information is that the prototype Taylor Cub was
first tried with a Tiger Jr. but that performance was
so unsatisfactory that the Cub design would have
been dropped had the Continental A-40 not come
along in the nick of time. This story is a bit strange,
since the Aeronca twin and the Szekeley radial
were available, but the Tiger trial seems well attested to. Some Tiger Kittens were sold in Oregon,
where the homebuilt movement survived the
dreary thirties. After the demise of the Brownback
firm, the Tiger Kitten was produced for a few years
by the Light Foundry Co. of Pennsylvania.

The longest-lasting American two-stroke engine was the Irwin, or Meteor. It was called a 4-
cylinder radial but was actually two opposed twins
at 90° to each other. It was designed by J. F. Irwin

of California, who set out to build an "everyman's
airplane" right after the First World War by simplifying the conventional biplane. Like many others,
he tried V-twin motorcycle engines with poor results
but, unlike most experimenters, went on to build
a pretty good engine of his own. The first Meteors
were very much country-shop products with boxy-
looking welded-on transfer passages and other
crudities, but they worked well enough. With the
passing of time they gained such refinements as
iron-lined aluminum cylinders and rotary valves.
The mature Irwin gave 25 hp at 2100 rpm and
claimed, no doubt optimistically, specific consumption of 0.50 lb (0.227 kg).

A similar-looking engine with the same name
was built in Italy during the fifties. The basic engine,
with crankcase compression and burning a fuel-
and-oil mixture, had specific consumption of 0.89
lb (0.4 kg) and was promoted as an engine for
target drones; on the supercharged and fuel-injected model, specific consumption dropped to half
that amount. This engine can have owed only its
conception and name to Irwin's Meteor; it was
considerably larger.

If any two-stroke engine had actually attained
the efficiency claimed for the Shackleton and the
Hurricane, the type would not have died out. The
difficulty with two-cycle engines may be a subtle
one. All power sources, whether steam, electric,
turbine, or internal combustion, must pay a price
in reduced efficiency when flexibility is a requirement; and two-strokes seem to carry a higher
penalty than other types. If a two-stroke engine
were optimized for one specific speed and load, it
might show a sensational reduction in specific consumption but would be a troublesome engine to
operate. Idling would be rough, part-throttle operation uncertain, and availability of extra power in
emergencies dubious.

Two-stroke engines continue to buzz through
the air today, but their use is confined to those
misguided people who fly for the fun of it. Hirths
and a few California-built Nelson Dragonflies (Fig.
11-8) haul self-launching sailplanes into the air.
Chain saw and snowmobile engines power the
foot-launched ultralights that, for the moment at
least, slip through the meshes of government reg-
ulations. Long may they live!

FIG. 11-8 Early post-war Nelson Dragonfly powered glider, 40-hp Nelson engine. The glider is based on Bowlus designs; later "self-launching" sailplanes have retractable engines. *(Courtesy of the author, taken at NASM-Silver Hill.)*

The Diesels[5]

Diesel engines use less fuel than any other kind of engine, they have no ignition systems to give trouble, and Diesel fuel is less flammable than gasoline. Unfortunately, their very high compression ratios demand great strength, which means weight, and their fuel is heavier than gasoline. Also, until the Germans developed "solid" fuel injection in the mid-twenties, they had to incorporate an air compressor and other complex accessories. As a result of these difficulties, aircraft "Diesels" before 1928 were not true Diesels. For example, the French Garuffa of 1921, a water-cooled 9-cylinder radial, was a supercharged two-stroke engine in which fuel was blown in with the scavenging air. It was a compression-ignition engine but lacked injectors. By the same token, the so-called Diesel model airplane engines are not real Diesels. This is not an academic distinction. Excess air is a major element in Diesel economy, since this guarantees that *all* the injected fuel is burned under *all* operating conditions.[6]

The first aircraft Diesel to achieve even partial success was the American Packard of 1928. This was a 9-cylinder, four-stroke radial whose weight

[5]A Diesel is rather strictly defined. It is a compression-ignition engine with fuel injection, burning a fuel heavier than gasoline.

[6]Truck mechanics, who sometimes set the injectors extra rich for more power, to the contrary.

and power closely matched that of the J-5 Whirlwind. It used only one valve per cylinder for both intake—of plain air, of course—and exhaust and, to save weight, had a magnesium crankcase to which the cylinders were held by steel hoops (Fig. 11-9). The Packard received considerable publicity and was awarded the Collier Trophy, but in the end it was a failure. Because of the single valve, there was no way to provide it with an exhaust manifold, and, as a result, the smell drove users out of their minds; Clarence Chamberlain said that he was never so happy to get rid of an airplane as he was with his Packard-powered Lockheed Vega. It vibrated so badly that it succeeded in doing what two generations of aerobatic pilots have been unable to do: it caused structural failures in the Waco Taperwing, that rock of an airplane. Its reliability, as a result of trying too hard to save weight and of putting it into production with inadequate development, was uncertain. Yet, to its credit, it established, in a Bellanca (what else?), a world's non-refueling endurance record of 84 ½ hr that stands to this day (Fig. 11-10).

Packards were tried out in a blimp, foreshadowing the most significant use of airborne Diesels. The firm had hoped to build a Diesel as light as a gasoline engine, and had this hope been realized, the Diesel would be practical in airplanes. If, however, a Diesel must be heavy to be reliable, its use would only make sense in a dirigible. The gasbags could be expected to stay aloft long enough for the Diesel's superior fuel economy to more than compensate for its greater weight, whereas airplanes could not.

Britain's major effort in a dirigible Diesel was the Beardmore. In retrospect, the Beardmore seems to have had everything possible wrong with it. To begin with, it was a straight eight, a configuration that almost always results in a heavy engine. Then, it was a four-stroke, rather than the generally better supercharged two-stroke. Instead of using unit injectors (pump and injector combined), it had pumps at the end of the engine with long tubes to carry the fuel to the injector nozzles, thus making the fuel flow subject to pulses and irregularities. A monster in size as well as in concept, it had a bore of 8¼ in. (210 mm) and a stroke of 12 in. (305 mm), giving a displacement of 5132 cu in. (84 L).

It gave 585 hp at 900 rpm and weighed 4000 lb (1712 kg), for the miserable weight-to-power ratio of 6.84 lb/hp (3.11 kg/hp). It was started by a small gasoline engine and required the full-time attendance of a crew member to keep running. The excess weight of the Beardmores contributed to the fiery demise of the R-101 on its first voyage. It was found necessary to overinflate the interior gasbags if the airship was to fly at all; they chafed against the structure and developed the leaks that caused it to crash into a hillside at Beauvais.

FIG. 11-9(a) Packard Diesel no. 1, the first Diesel ever to fly in a heavier-than-air machine. *(Courtesy of the author, taken at NASM-Silver Hill.)*

The U.S. Navy also put some money into a dirigible Diesel, the Attendu, in 1925. This was a two-stroke engine with the feature of valve-throttled exhaust ports to promote better slow running. In this connection, it is a fact that some Diesels, when well broken in, do not stop running even when the fuel is completely shut off! This is because the lubricating oil left on the cylinder walls ignites from

This lever must be positioned so as to allow .010 to .015 end play of tube

FIG. 11-9(b) Cross section of the Packard Diesel. Note the little cam followers and pushrods to operate the unit-type injectors. In most other respects, the construction follows conventional radial engine practice. *(Courtesy of NASM-Smithsonian.)*

FIG. 11-10   One day in 1931, Packard test pilots Walter E. Lees (in cabin) and Frederic A. Brossy loaded this Bellanca up with fuel and took off. They landed 3 days, 12 hours, and 33 min later, finally out of fuel! *(Courtesy of NASM-Smithsonian.)*

the compression and keeps them turning over. Before the Attendu was fully developed, the *Los Angeles* entered service and the excellent performance of its Maybach gasoline engines led the Navy to decide that it had better things to do with its limited funds.

Another promising dirigible Diesel came, somewhat oddly, from Belgium. This was the big (1200-hp) Deschamps, a two-stroke V-12. It was reversible for maneuvering, and there was some talk of using this feature in an airplane for reverse-thrust (what is called today "Beta mode") rapid descents. Unfortunately for Deschamps, the Germans were by then the only ones persevering with the big gasbags and they had their own good engine sources.

The best known of these was Maybach, which built mostly gasoline engines but did develop one Diesel, its type GO 56. Its maximum power was not announced, but it had a continuous rating of 410 hp. Since it weighed 4620 lb (2100 kg) dry, its power-to-weight ratio was about as dismal as the Beardmore's. Perhaps because of this weight, the *Graf Zeppelin* was not Diesel powered, using instead Maybach gasoline engines converted to

burn "Blau gas," a gaseous fuel whose weight was the same as that of air. These Maybachs that powered the *Graf Zeppelin,* and the American *Akron* and *Macon,* were big [2024-cu-in. (33.2-L)] V-12s in the old separate-cylinder tradition with the added feature of being reversible. This was accomplished by having the camshaft arranged to slide axially, engaging a different set of cams to time the valves for opposite rotation. A safety linkage was incorporated to assure that reversing could not be attempted until the ignition was switched off and the throttle was closed.

Daimler-Benz built the largest of all aircraft Diesels, its type LOF 6, for the *Hindenburg*. This company had been gradually working its way back into the aircraft engine business since dropping out in 1919 and, in addition, had produced some large truck and marine Diesels. As a result of their experience, the four-stroke LOF 6 was apparently an entirely satisfactory engine. It is no great trick to bring a high-production engine to a state of acceptable reliability, since experience accumulates quickly and debugging can be rapid; nor is it hard to build a reliable engine if it is worked at a speed and load well below its potential maximum. What *is* a real achievement is making a reliable semi-custom engine of high output. Probably the Rolls-Royce R Schneider Cup engine was the outstanding example of this, but if so, the LOF 6 was a close second.

It was a V-16 that incorporated some of the design features of the earlier F-2 gasoline V-12. At its maximum output, 1200 hp, its specific consumption was 0.39 lb (0.177 kg), and at 900-hp cruise power specific consumption was 0.37 lb (0.168 kg), as economical as any engine that ever flew. This big engine weighed about 4400 lb (2000 kg), giving a weight-to-power ratio of 3.67 lb/hp (1.67 kg/hp), excellent for a Diesel. Clearly this was not an engine that was derated or overly conservatively designed in the interest of reliability. Incidentally, an idea of the magnitude of this last of the Zeppelins can be gained from the fact that even with the economical LOF 6, the *Hindenburg* required a fuel tank capacity of 66 tons (60,000 kg).

Most major engine-building nations experimented with Diesels during the late twenties, but, compared with the daring of Packard and the solid achievements of Mercedes and of Junkers (to be described), their efforts amounted to little. Rolls-Royce, for example, modified a Condor into a Diesel. Rather than design a new lower end to take the Diesel stresses, Rolls reduced the bore and increased the stroke of the standard engine. Unit-type injectors were used. The engine passed its official tests at 500 hp in 1934, and the weight-to-power ratio was excellent at just under 3 lb/hp (1.36 kg/hp). However, whether because of the abandonment of the dirigible program or for some other reason, it was never put into production.

There were two other abortive British Diesels. One, the Napier Culverin, was simply a license-built Junkers. The other was the Coatalen, designed by the talented Swiss engineer who had been responsible for much of the Sunbeam car's success. A liquid-cooled V-12 of 500 hp, it had no chance against the radials that by then dominated its power class.

Except in Germany, that is. While most of the world was beginning to follow the trail blazed by the American DC-2, Boeing 247, and Lockheed 10, Germany was taking a different path. The most prominent German designers, men of the ilk of

Heinkel, Messerschmitt, and Tank, were happy to do all their advanced work on aircraft for military use, while old Professor Junkers had been squeezed out early by the Nazis. There was no official interest, the only kind that counted, in advanced civil aircraft unless they had the potential for being converted into bombers. As a result, the German transport planes didn't need 1936-type engine performance because they were 1932-type airplanes. Where the high wing loading of a DC-3 required the full 5-min-rated 1000 hp of its engines on takeoff, a Ju 52 3m had only 75 percent as high a wing loading and could float off the runway easily on something close to its cruising power. Thus, a good aircraft Diesel, with its appealing fuel economy, was bound to meet more acceptance in Germany than anywhere else.

The great aircraft Diesel was, of course, the Junkers. Its origin went back to the U-boats of World War I. Their Diesels were inline sixes with opposed pistons and two crankshafts, one below and one on top. This construction gives the same displacement as a V-12 with less width, or the same as a straight eight with less length and weight. In addition, vibration is probably less than with any other type of Diesel. Junkers may have had some connection with these marine engines, but this is uncertain; in any case, when he turned his hand to building an aircraft Diesel, he followed the same pattern. The first Junkers Diesel, a liquid-cooled engine very much like the U-boat engines in design, flew in 1929. By 1932 the Jumo Diesel was sufficiently developed to be continued with only minor changes through 1944.

The opposed-piston concept was very well exploited in the construction of the Junkers aircraft engines. A single cast-aluminum block extended from slightly above the centerline of the upper crankshaft to slightly below that of the lower one. Sturdy cast covers, held by a multiplicity of bolts, contributed to the general rigidity. In the absence of cylinder heads, the steel cylinder liners were easy to insert from one end and, the engine being a two-stroke, there were no complexities resulting from the need to drive camshafts. Air was blown in at the top, fuel was injected in the middle, and the exhaust came out at the bottom.

The classic Jumo Diesel was the 205, which

gave up to 560 hp, weighed 1257 lb (571 kg), with a specific consumption of between 0.375 and 0.396 lb (0.17 to 0.18 kg). A bit tall, narrow, and fairly short, it was a neat and professional-looking design. The 205's displacement was 1014 cu in. (16.62 L), the Jumo 206 gave 1050 hp from 1525 cu in. (25 L), and the 207, a turbocharged 205, gave up to 1000 hp and had enhanced high-altitude performance (Fig. 11-11).

Enough of these engines were used by Lufthansa in passenger service for some interesting statistics on safety to be compiled. It was found that there was no significant difference in accident frequency rates between gasoline- and Diesel-powered planes, but that survival rates in Diesel crashes were considerably higher. The crashes of gasoline-powered planes tended to be accompanied by fire; those involving Diesels did not.

Perhaps the Diesel should have had a post-war history. Its combination of safety and economy would have been ideal for such long transport routes as TWA's Paris–New York and SAS's transpolar route. On such flights, the airplane is carrying an enormous load of fuel at takeoff, and had this fuel been Diesel oil instead of gasoline, the risk would have been substantially reduced. The extra weight of Diesels large enough to get a DC7-C or

a Lockheed Super-Connie into the air would have been more than made up by the reduced weight of fuel resulting from the Diesel's economy at cruise settings.

With this in mind, Napier put its money into a fantastic post-war powerplant called the Nomad. If other engine builders accepted complexity, when necessary as a means to an end, Napier seemed almost to embrace it as an end in itself. The Nomad was a two-stroke opposed 12 Diesel turbo-compound with added jet thrust! It was the ultimate example of trying to touch all bases. Its specifications were impressive. Including a component derived from converting its 250 lb (113 kg) of jet thrust to horsepower, the Nomad gave 2248 continuous hp at a specific consumption of 0.33 lb (0.15 kg). Its rated maximum was 3570 hp, which may have included the extra thrust from a jet afterburner that was part of the package (Fig. 11-12).

But time was running out for the big piston engines. The future clearly lay with the turbine. The turboprop Vickers Viscount was flying by 1948, and the Comet, the first jet transport, entered service in 1952. In the interim the well-proven Bristol, Wright, and Pratt and Whitney radials would do. The Wright Turbo-Compound, especially, giving (sometimes) the fine specific consumption of 0.38 lb (0.17 kg) could be used in existing airframes and looked like a low-risk long-range engine compared with the new and untried Nomad. The Nomad was a failure on the market, and Napier turned its talents to the development of the equally complicated Deltic Diesel for railroad use.[7]

Automobile Conversions

Automobile engines are designed very differently from those for airplanes and do not ever make really good aircraft powerplants. However, they are cheap and sturdy, and so there has always been some temptation to try to adapt them for aircraft. The most successful such conversion was probably B. H. Pietenpol's rework of the Model A Ford in

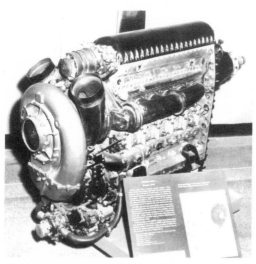

FIG. 11-11 Junkers Jumo 207 Diesel, 1000 hp, turbosupercharged. (Courtesy of NASM-Smithsonian.)

[7]The Deltic has six pistons per row and three crankshafts.

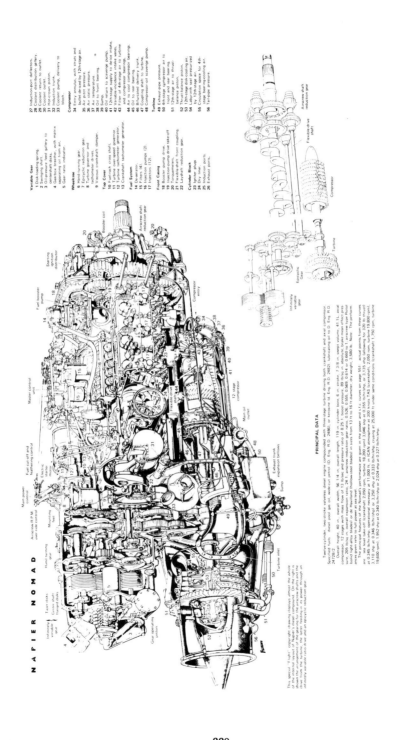

NAPIER NOMAD

Variable Gear
1 Disk-loading spring.
2 Swinging shaft.
3 Oil-pressure feed gallery to centershaft disks.
4 Gearbox breather, with matrix separating oil from air.
5 Gear ratio indicator.

Wheelcase
6 Hand-turning gear.
7 Epicyclic reduction gear.
8 Turbine governor and tachometer drives.
9 Viscous crankshaft damper.

Top Cover
10 Fuel-rack cross shaft.
11 Turbine over-speed governor.
12 Turbine tachometer generator.
13 Crankshaft tachometer generator.

Fuel System
14 De-aerator.
15 Filters (4).
16 Injection pumps (2).
17 Injectors (12).

Variable Gear
27 Induction-gear deflectors.
28 Coolant distribution gallery to block.
29 Coolant return to outlet.
30 Coolant gallery.
31 Hot-crown piston.
32 Induction trunk.
33 Coolant pump, delivery to block.

Compressor
34 Entry annulus, with struts and bullet de-icing by 12th-stage air.
35 Air pitot pressure.
36 Air static pressure.
37 Air temperature.
38 Oil to front bearing.
39 Sump.
40 Oil return to scavenge pump.
41 12th-stage supply to de-icer intake.
42 Variable-incidence intake vanes.
43 Filter of 4th-stage air to turbine and compressor bearings.
44 Air to cool compressor bearings.
45 Oil to rear bearing.
46 Bifurcated delivery trunk.
47 Coupling shaft to turbine.
48 Compressor-oil scavenge pump.

Turbine
49 Exhaust-pipe pressure.
50 4th-stage compressor air to front and rear bearing.
51 12th-stage air to thrust-balance piston.
52 Thrust-balance piston.
53 12th-stage disk-cooling air.
54 Labyrinth seal, pressurized by 12th-stage air.
55 Circulation space for 4th-stage bearing-cooling air.
56 Tail-pipe pressure.

Front Casing
18 Booster pump drive.
19 Injection-pump drive take-off.
20 Torquemeters.
21 Flexible-shaft front coupling.
22 Layshaft reduction gear.

Cylinder Block
23 Igniter plugs.
24 Dry liner.
25 Induction ports.
26 Exhaust ports.

Airscrew-shaft reduction gear

Flexible-drive shaft

Compressor

Turbine

Epicyclic gear

Infinitely variable gear

Main power control

Airscrew R.P.M. over-rate control

Fuel cut off and feathering control

Mounting feet

Electric starter motor

Hand-turning gear

Cross-shaft forward disks

Con-rod disks

Cam disks

Infinitely variable gear

Gear operating piston

Starting injection distributor

Booster coil

Airscrew shaft reduction gear

Fuel booster pump

Master control unit

12 stage compressor

Main oil outlet

Exhaust trunk expansion bellows

Sump

Turbine inlet

This special "Light" copyright drawing displays almost the whole of the internal layout of Napier's diesel compound. On the right is shown the arrangement of the gearing for the airscrew shafts and the drive from the turbine, the latter feeding in its power through an infinitely variable ratio drive and an epicyclic reduction gear.

PRINCIPAL DATA

Twelve-cylinder, two-stroke valveless diesel engine compounded with three-stage turbine driving both crankshaft and axial compressor
Specified fuels: diesel (pool gas oil, wide-cut petrol (D. Eng. R.D. 2486), or kerosene (d. Eng. R.D. 2482), lubricating oil to D. Eng. R.D. 2472B/2

Overall height, 40 in., overall width, 56 1/4 in, overall length, 119 in, cylinder bore, 6 in, stroke, 7 3/8 in, swept volume, 41 L, axial compressor, 12 stages with mass flow of 13 lb/sec and pressure ratio of 8.25:1, boost pressure, 89 lb/sq in absolute, brake mean effect pressure, 205 lb/sq in, overall expansion ratio, 24:1, airscrew reduction gear ratio, 0.526, 0.565, 0.569, 0.614, or 0.660 to 1, airscrew type Rotol (solid light-alloy blades) or de Havilland (hollow-steel blades) in turn from 13 ft to 16 ft diameter, dry weight, 3,580 lb. Note: The performances given refer to full power at sea level.

The principal features of the Nomad's performance are given in the power and t.c. curves on page 551; actual points from these curves are: sea level take-off (crankshaft 2,050 rpm, turbine 18,200 rpm) 3,046 shp at 0.355 lb/hr/shp, or 3,135 shp (allowing for 320 lb thrust at 0.345 lb/hr/shp operational necessary at 11,000 ft, in ICAN atmosphere at 300 knots TAS (crankshaft 2,050 rpm, turbine 19,800 rpm), 3,110 shp at 0.346 lb/hr/shp or 1,250 shp at 0.333 lb/hr/shp, cruising at 25,000 ft under same conditions (crankshaft 1,750 rpm, turbine 19,600 rpm) 1,952 shp at 0.340 lb/hr/shp or 2,024 shp at 0.327 lb/hr/shp.

FIG. 11-12 Napier Nomad. The reader may consider this a complex engine, but Napier did not; for example, it described the flat-twelve two-stroke Diesel that formed its reciprocating component as "the simplest possible type." (*Courtesy of Flight via NASM-Smithsonian.*)

1930. The Model A had its faults, but if a great engine is one that has successfully powered cars, trucks, racing cars, and aircraft, the Model A qualifies. Pietenpol turned it back to front, changed it over to pressure oiling instead of splash and to magneto ignition, made a few other minor modifications, and had an airplane engine. Running at 1600 rpm, it produced only about 23 hp; but its 118 ft-lb (160 N·m) of torque swung a big propeller from a World War I Lawrance and hauled the 1010-lb (460-kg), aerodynamically dirty Air Camper along at 75 mph (120 km/hr). A later, commercially made, certificated conversion of the Model A, of whose details I have no knowledge, went into the two-seat "Wiley Post" biplane of the mid-thirties and apparently gave satisfaction; but by then low-priced opposed engines were available. (I use quotation marks because Wiley Post had nothing to do with it except for permitting the use of his name.) (Fig. 11-13.)

The Model A had a small cousin in the English Ford 8 of 57 cu in. (993 cc). This was inadequate for any normal airplane, but when M. Henri Mignet's *Pou de Ciel* came out, the Cardan Engineering conversion of this little four went into

FIG. 11-13 Aircraft conversion of a Model A Ford, once used in a Pietenpol Air Camper. The radiator position is authentic; the pilot had to peer around it in flight. (*Courtesy of the author, taken at Glenn Curtiss Museum, Hammondsport, New York.*)

several of the English-built examples. It produced enough power, but its weight upset the *Pou's* somewhat critical longitudinal stability, and a number of deaths resulted, after which the Air Ministry banned all "Fleas"—not just the nose-heavy ones.

Pietenpol made an even more unlikely conversion of a Ford motor the next year, when he flew a single-seat version of his Air Camper, called the Sky Scout, behind a Model T. This very old 180-cu-in. (3-L) needed more modifications than the Model A,[8] and while it worked well enough—everything Pietenpol did worked—it was never popular. Pietenpol probably did it to see whether he could succeed where many others had failed rather than with any idea that it would achieve wide popularity. In any case, the whole question became moot when the government passed some new rules that wiped out the homebuilt movement for the next 20 years.

Automobile engines of later design were less well suited for aircraft conversions because they were intended for higher compression ratios and higher sustained speeds. This made them heavier per unit of displacement than such older types as the Model A. Nevertheless a few conversions of later motors were made. Most of these in the United States were under the auspices of a program sponsored by Eugene Vidal, Assistant Secretary of Commerce for Air, in 1936. With $1 million of federal money to spend, Vidal tried to find out whether a plane could be built that would cost no more than a medium-priced car. Most of the entries in the government's contest were automobile-engine-powered. Ole Fahlin, the propeller maker, built one with a Plymouth six, Arrow used a Ford V-8, and Waldo Waterman put a Studebaker six in his tailless "Arrowbile." None was really marketable, though the Arrow saw limited commercial production. The reason for this is simple. To fly with an overweight engine, the airframe has to be bigger and the added cost of building the larger airplane wipes out any saving in engine cost. One private effort along the same lines was made just before the Second World War: the Akron Funk,

[8]Pietenpol said it had a bent hairpin for a crankshaft.

which was produced with a converted Ford four. When later equipped with an 85-hp airplane engine, the Funk turned out to be a very good airplane; but it was a hopeless proposition with the Ford.

A few homebuilt conversions, all characterized by the use of reduction gears, have flown in recent years. One was an aluminum-block Oldsmobile V-8, which was said to give good performance in a Cessna, but nothing further has been heard from it. Another was based on the seven-main-bearing, 176-cu-in. (2.74-L) Ford Falcon six; Don Stewart, of Stewart Headwind fame, flew it in a modernized Lincoln Sport that he called the Foo Fighter, but he reported the need for further development that he wasn't prepared to undertake. At present, the conversion of water-cooled automobile engines seems to be about finished.

Air-cooled car engines, notably Volkswagen conversions, are more popular. They vary greatly in quality. Many amateur conversions are decidedly lethal. People have drilled holes in the heads to convert them to dual ignition, made strange manifolds for updraft carburetors, built peculiar reduction gears, and in general have operated on the Orville Hickman level. They seem to know nothing about how Volkswagen engines are made. For example, after one propeller hub came off in flight, with fatal results, it was learned that the only thing holding a VW flywheel in place is a press fit on a taper at the end of the crankshaft. On the other hand, a first-class conversion, such as a Revmaster, is a quite satisfactory engine in an aircraft designed with the VW characteristics in mind.

The first Volkswagen conversions were made in France around 1950. Those early VWs had a displacement of 67 cu in. (1100 cc) and gave 27 hp at something over 3000 rpm. In efficient single seaters like the Jodel Bébé with a 24-ft (7.3-m) span and with a 700-lb (318-kg) gross weight, and the hardly less efficient Druine Turbulent, they operated rather satisfactorily under French conditions. These engines had no reserve power for takeoff and climb was sluggish but this was not a serious drawback when operating out of large abandoned military fields with light traffic.

The trouble, especially in America, came when displacement and power began to rise. The 77-cu in. (1250-cc) VW was still conservatively rated but the later and larger engines developed their power at higher rpms. Evidently people expected too much from them, not realizing that the higher speed meant that thrust did not rise proportionately with power. The resulting disappointing performance led to some very ill-advised modifications that eroded the engine's reserve of strength, and reliability naturally suffered.

The Corvair attracted considerable attention for a while. Here the high revs. were recognized early as a potential problem and some Corvair conversions used reduction gears. Some direct-drive conversions were made, but these were successful only where it was recognized that the Corvair had the same displacement as a 65-hp aircraft engine and ought not to be asked for more thrust than such a powerplant would deliver. B. H. Pietenpol put a practically stock Corvair in a J-3 Cub and got just about the same performance from it that the A-65 gave: more ambitious installations were less satisfactory. Pietenpol's Corvair ended up in an Air Camper, where it proved to be a satisfactory replacement for a Model A Ford!

Like the Model A, the Volkswagen is a great engine in its way. But a good one isn't cheap. The old idea of getting an engine from a low-mileage wreck and putting it directly into a nice little plane is largely out of the window unless you fly out of a low-traffic airport in flat country. A proper conversion costs money. But, in a sense, you may have no choice, because the A-65s are getting worn out and there are no other small engines in production.

The Giants

The U.S. and, to a lesser extent, the British and German governments underwrote the development of some enormous engines at the time of the Second World War. The best of these, and the only ones to reach service status, were actually doublings of existing engines. The German D-B 610, mentioned earlier, was made from a pair of D-B 605s built together. The Rolls-Royce Vulture was a 24-cylinder X-type, cobbled together by hooking up a pair of Kestrels to a common crankshaft. It gave 1845 hp but never got the development work needed to make it a reliable engine; better than the D-B 610, it was nevertheless a failure. The Vulture was adopted as the powerplant for the Avro Manchester at the beginning of the Second World War. The Manchester was a twin-engine heavy bomber that perhaps showed what the USAAF had in mind for the large engines it was sponsoring: a "heavy" that didn't need four engines. After numerous in-flight fires forced the Manchester's withdrawal, it was redesigned, in one of the war's best demonstrations of British improvisatory talent, into the four-Merlin Lancaster—the war's best bomber in the 4800-hp class. (Apologies to B-17 veterans, but isn't it true that a loaded Lanc could outrun an empty B-17?)

The American double engine was the Allison V-3420. It was begun as early as 1937 and was, in a sense, the least technically ambitious of the lot. It also might have been the best of the three had it been developed. It was simply two V-1710s, each with its own crankshaft, spread out fanwise on a common crankcase and geared to a common propeller shaft. When completed in 1940 it was the world's most powerful aircraft engine at 3000 hp. It did, however, have one insoluble problem: there was no airplane that it could be used in (Fig. 11-14).

Other American giants were made by Pratt and Whitney, Continental, Lycoming, Chrysler, and, possibly, Studebaker. The Studebaker engine was probably never built, but if it was, it was the biggest aircraft engine. It was to have been a 24-cylinder H-type capable of developing 5000 hp and would have had the incredible displacement of 9350 cu in. (153 L). Most of the American giants followed the same H-type configuration, but the biggest one known to have been actually built, the Lycoming XR 7755 (127 L), was a "corncob" engine. The name came from the idea that the 36 cylinders, arranged in four rows of 9 cylinders, resembled the kernels on an ear of corn. I have seen this engine, and its dimensions are, like those of a Bugatti Royale, almost incredible. It developed 5000 hp on test, and the target was 7000. There were nine overhead camshafts which could be shifted axially for METO power in one position and cruise power in the other. Two great shafts emerge for coaxial propellers, and there was a two-speed gear-change box between the crankshaft and the propeller shafts. At this writing, it resides in a crate at Silver Hill. Everything I have ever said about misdirected effort in any nation fades into insignificance next to this monumental American folly. I have read a lot of aviation material and talked to a lot of aviation people in the 35 years since this 3-t (13,200-kg) monstrosity was built and have yet to hear the least hint as to what sort of airplane the government thought it could go into.

Chrysler's 1800-hp IV 2220 was a somewhat

FIG. 11-14   Allison V 3420, 24 cylinders, 3000 hp. Why has this prototype got two propeller couplings? Could it have been intended for marine use? *(Courtesy of Detroit Diesel Allison Division, General Motors Corporation.)*

more conventional giant. It was a V-16 that has been described as two V-8s built together, but, in my opinion, it was too well integrated as a V-16 for this to be valid. It is true that the driving gears for its propeller shaft were in the middle, but Mercedes-Benz built a post-war straight eight Grand Pix car this way without anyone calling it a pair of fours. Driving this way makes considerable sense because it avoids the possibility of damaging torsional resonance—always a likelihood in a crankshaft that is eight cranks long (Fig. 11-15).

Pratt and Whitney's H-3130 was an outgrowth of its similar but smaller X-1800. Both were 24-cylinder H-type sleeve-valve engines with separate cylinders rather than cast blocks. The separate cylinders were selected for ease of servicing, and the sleeve valves resulted from Pratt and Whitney Chief Engineer George Mead's friendship with Roy Fedden. Output figures for both engines were very impressive; they were well into the Hyper class. The X-1800 gave 2000 hp from 2599 cu in. (42.6 L) and weighed 2400 lb (1091 kg). It could be operated either flat or vertical. It was nearly 9 ft (2.71 m) long, but its minimum dimension was 24 in. (610 mm) wide if operated vertically or 15 in. (380 mm) high if horizontal. The distance across the cylinders, at 51 3/16 in. (13,000 mm) was less compact; as with the Continental mentioned in Chap. 5, it would have been at its best in a buried installation. The H-3130 was simply the same engine scaled up, having 3730-cu.-in. (61-L) displacement, 2650 hp, and a weight of 3350 lb (1523 kg). Leading dimensions went up proportionately. After getting these projects started, Mead was replaced as chief engineer by Hobbs, a nonbeliever in liquid cooling, and both projects lost their momentum. Finally, in 1941, Rentschler told General Arnold that Pratt and Whitney could develop the R-4360 or the H-3130 but not both. Arnold agreed with Rentschler and Hobbs that the Wasp Major was the engine that Pratt and Whitney should concentrate on, and the Hypers were cleared off by 1943 (Fig. 11-16).

All these engines, except the Allison and the Chrysler, were liquid-cooled engines built by firms that had specialized for years in air cooling. This was inevitable, since Allison was the only American builder of liquid-cooled aircraft engines after 1934, but one giant did reflect its makers' devotion to air

FIG. 11-15 Chrysler IV 2220. This display exhibit was intended to rotate, in order to allow visitors a better look. Does the I designation mean that it was intended to be an inverted engine? *(Courtesy of the author, taken at Bradley Air Museum, Windsor Locks, Connecticut.)*

cooling: Wright's 2160-cu.-in. (39.4-L) 42-cylinder corncob (barely a giant but in the Hyper class). As it originally emerged, it had air cooling for the front cylinders and liquid cooling for the rest! When it was actually built and run, cooler heads prevailed and it was entirely liquid-cooled.

The almost universal acceptance of the sleeve valve in these big engines is noteworthy. Rolls-Royce accepted the fact that the poppet valve was not as suitable for giant engines as early as 1936, when it began the design of a 24-cylinder, 3600-cu.-in. (22-L) X-type, cleverly named the Exe. The Exe was dropped after 1938, but sleeve valves were used in Rolls' last and biggest piston engine, the Eagle II. The Eagle was begun as late as 1942, as stated in Chap. 5. This late beginning meant that the whole project was a grievous loss to a country that could hardly afford such mistakes; and it was yet another H-24. It was rather like a Sabre scaled up to Exe size. The fate of the 3500-hp Eagle II is historically significant. Driving counterrotating propellers, it was intended to power the striking Westland Wyvern carrier-based fighter-bomber. But the Wyvern became a victim of the characteristic British post-war on-again, off-again hesitation in aerospace projects and, by the time it was ready, so was the Trent turboprop; the Eagle never went into service,

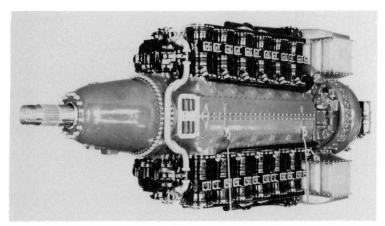

FIG. 11-16   Pratt and Whitney H-3130. The Pratt and Whitney "Hypers" were actually the first of the sleeve-valve H-24s to be built, but the United States abandoned the concept and only the British Napier Sabre saw service in World War II. *(Courtesy of Pratt and Whitney Division, UTC Corporation.)*

and the Wyvern became the world's only production turboprop-powered fighter.

And One More

This is a history of piston engines, and not all piston engines have been internal-combustion types. The gasoline engine made the airplane possible, but steam has always had its advocates, and, in 1930, the steam engine finally powered an airplane. Besler, a maker of logging locomotives, installed a V-twin compound in a Waco Ten (some sources say a Travelair) and flew successfully in California. Steam enthusiasts, as might be expected, went bananas at this, but Besler broke their hearts by announcing that he had no intention of trying to develop his engine further; he had proved that it could be done and was satisfied (Fig. 11-17).

Lessons

I think that there is little question but that, in this compendium of lost causes, there are some ideas that could have been developed into popular engines. Some of them didn't make it because of inadequate financing and some because of unfortunate timing. But most of the failures came from attempting too much. When someone saw a way

to try one improvement at a time, as Fedden did in putting sleeve valves in an otherwise conventional radial, success often followed. It has been the attempt to go back to first principles and start with a fresh sheet of paper that has characterized most of the failures. If you have one earthshaking idea that can be used with known technology or every part of your engine except for the new thing, you have a chance; but if you try two ideas at once, you are probably doomed from the start.

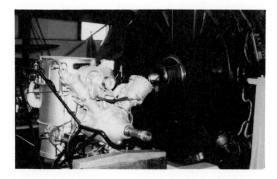

FIG. 11-17   Besler's V-twin compound: the only steam engine to ever power a man-carrying heavier-than-air craft (replica). Besler's logic in trying it in an airplane built for an OX-5 was excellent; an airplane that could fly decently with an OX-5 ought to be able to fly with anything. *(Courtesy of the author, taken at NASM-Silver Hill.)*

Data for Engines

Maker's name	Date	Cylinders	Configuration	Horsepower	Revolutions per minute	Bore and stroke, in. (mm)	Displacement, cu in. (L)	Weight, lb (kg)	Remarks
Alfaro	38	8	NA	155	2000	2.8 (71) × 3.4 (87)	167 (2.7)	240 (109)	Power is clearly very optimistic
Statax	14	5	Ro, B, A	40	1200	3.9 (100) × 4.7 (120)	288 (4.7)	200 (91)	
Macomber	11	7	Ro, B, A	50	1000	7 (178) × 4.3 (108)	1145 (18.8)	230 (105)	Bore and displacement seem too large; possibly 7 in. (178 mm) should be stroke
Redrup	NA	7	B, A	95	2200	NA	NA	200 (91)	
Fury	39	5	B, A	NA	NA	3.4 (114) × 5 (127)	232 (3.8)	160 (73)	
Gadoux	38	12	B, L	900	3600	6.3 (160) × 4.3 (110)	1555 (25.5)	NA	Actually six opposed twins; had three-throw cranks like experimental Anzani
Cleveland model 4	NA	6	B, L	150	NA	5 (127) × 6 (152)	707 (11.6)	NA	
Fairchild-Caminez	25	4	R, A	135	1000	5.6 (144) × 4.5 (114)	447 (7.3)	360 (164)	Two piston strokes per revolution
Gobron-Brillé	10	4–8	R, L	102	1400	4.7 (120) × 7.9 (200)	1102 (18.1)	350 (159)	Combined stroke of the opposed pistons was 400 mm
Dawn	30	5–10	R, A	156	2500	4.1 (105) × 3.8 (95)	501	NA	As Gobron-Brillé
Beltrame	NA	2	I, A	18	NA	NA	NA	NA	
Sega	NA	7	Ro, A	NA	NA	NA	NA	NA	
Edelweiss	13	6	Ro, A	NA	NA	NA	NA	NA	
Deeble Duplex	37	6	Iv, I, A	300	2000	4.5 (114) and 4.3 (110) × 5 (127)	447 and 441 (7.8) and (7.2)	918 (417)	"Push" side of cylinders had different bore from "pull" side
Hermann	25	7	Ro, A	53	NA	3.4 (86) × 4 (102)	251 (4.1)	150 (68)	
Mawen	38	7	Ro, A	50	NA	2.7 (68) × 2.8 (70)	109 (1.8)	NA	

(cont.)

Data for Engines (continued)

Maker's name	Date	Cylinders	Configuration	Horsepower	Revolutions per minute	Bore and stroke, in. (mm)	Displacement, cu in. (L)	Weight, lb (kg)	Remarks
Bakewell Wingfoot	NA	8	R, A	165	2800	4 (102) × 4.5 (114)	452 (7.4)	380 (173)	Geared to 1800 rpm at propeller
Universal	NA	8	R, A	400	3000	NA	NA	NA	Geared to 1500 rpm at propeller
Jalbert	37	6	I, L	180	1930	4.9 (125) × 7.1 (180)	772 (12.7)	NA	Figures apply to Diesel version of Jalbert
Grade	32	4	Iv, I, A, TS	30	1800	2.6 (65) × 3.9 (100)	81 (1.3)	66 (30)	
Diecke	37	2	O, A, TS	18	2600	3.1 (78) × 3.1 (78)	46 (0.76)	62 (28)	
Scott Flying Squirrel	35	2	Iv, I, A, TS	16	2800	2.9 (73) × 3.1 (78)	42 (0.69)	85 (39)	
Aubier et Dunne	35	2	Iv, I, A TS	17	4000	2.8 (70) × 2.8 (70)	33 (0.54)	NA	Geared to 1600 at propeller
DKW	32	2	Iv, I, L, TS	21	4000	2.9 (74) × 2.7 (68)	36 (0.59)	82 (37)	
AVA type 4A00	35	4	O, A, TS	28	2800	2.8 (70) × 2.8 (70)	66 (1.1)	82 (37)	
Von Festenberg-Packisch	32	4	R, A, TS	40	NA	NA	NA	136 (62)	
Schliha F-1200	NA	2	O, A, TS	36	3100	3.7 (94) × 3.9 (100)	85 (1.4)	NA	Whether displacement includes a deduction for the piston extensions is not known
Caunter	NA	5	R, A, TS	60	2000	3.3 (85) × 3.5 (88)	153 (2.5)	160 (73)	
Shackleton model C	38	4	Iv, I, A, TS	60	2706	3.2 (82) × 3.9 (100)	129 (2.1)	120 (55)	
Wichita Blue Streak	29	5–10	R, A, TS	125	1800	2.2 (57) and 3.8 (95) × 3.8 (95)	250 (4.1)	125 (57)	Five sets of paired cylinders, each pair having one large bore and one small bore cylinder
Hurricane	29	8	R, A, TS	150	1800	4.5 (114) × 3.5 (89)	450 (7.4)	225 (102)	Rpm for propeller; engine rpm NA

Name		No. cyl.	Config		rpm	Bore × stroke			Remarks
Brownback Tiger Kitten 30	32	2	Iv, I, A, TS	30	1500	3.8 (95) × 3.5 (89)	77 (1.3)	75 (34)	
Irwin Meteor	29	4	R, A, TS	20	1730	3.1 (79) × 2.8 (70)	84 (1.4)	58 (26)	Figures for developed model
Meteor (Italian) G90 CA	59	4	R, A, TS	90	2800	4.1 (105) × 3.1 (78)	165 (2.7)	121 (55)	
Nelson H-63-CP	62	4	O, A, TS	48	4400	2.7 (68) × 2.8 (70)	63 (1)	68 (31)	Used 42-in. (1067-mm)-diameter propeller, negating the fine power-to-weight ratio
Packard DR-980	28	9	R, A	225	1950	4.8 (122) × 6 (152)	982 (16.1)	510 (232)	
Beardmore Tornado	30	8	I, L	585	900	8.3 (210) × 12 (305)	5132 (84.1)	4179 (1900)	
Attendu	25	2	I, L	85	1620	5.5 (140) × 6.5 (165)	310 (5.1)	500 (227)	Weight was above, and power below, expectations
Deschamps	34	12	Iv, V, L	1200	1600	6 (152) × 9 (229)	3052 (50)	2400 (1091)	Weight, if accurate, very low for a Diesel of this size
Maybach GO 56	NA	12	V, L	410	1400	5.5 (140) × 6.9 (175)	1973 (32.3)	4620 (2100)	Compare weight and displacement with Deschamps
Maybach VL-2	28	12	V, L	570	1600	5.5 (140) × 7.1 (180)	2029 (33.3)	2530 (1150)	Graf Zeppelin's engine; used gaseous fuel
Daimler-Benz (Mercedes) LOF 6	32	12	V, L	1200	1750(?)	6.5 (165) × 8.3 (210)	3299 (54.1)	4400 (2000)	Hindenburg's engine
Rolls-Royce Condor Diesel	28	12	V, L	500	1900	5.5 (140) × 7.5 (191)	2138 (35)	1504 (684)	This is same bore and stroke as gasoline Condor; some sources say bore was less, stroke more
Coatalen	36	12	V, L	550	2000	5.9 (150) × 6.7 (170)	2196 (36)	1210 (546)	
Junkers Jumo 205	32	6–12	I, L	600	2200	4.1 (105) × 2 × 6.3 (160)	1012 (16.6)	1147 (521)	Jumo 207 was turbosupercharged 205
Junkers Jumo 206	36	6–12	I, L	1050	NA	NA	1526 (25)	NA	
Napier Nomad	50	12	O, L	3750	NA	6 (152) × 7.4 (187)	2510 (41.1)	4200 (1907)	Hp as such not significant because of jet thrust component
Pietenpol	31	4	I, L	35	1600	3.9 (78) × 4.3 (108)	201 (3.1)	244 (111)	Hp peaks at 39 at 2200 rpm; torque peaks at 121 at 1000 rpm
Fahlin	37	6	I, L	80–90	3600	NA	NA	NA	

(cont.)

Data for Engines (continued)

Maker's name	Date	Cyl-inders	Config-uration	Horse-power	Revolutions per minute	Bore and stroke, in. (mm)	Displace-ment, cu in. (L)	Weight, lb (kg)	Remarks
Arrow	38	8	V, L	82	3075	3.1 (78) × 3.8 (95)	221 (3.6)	402 (183)	Note that weight per cubic inch is greater than Pietenpol and Funk "model A" conversions
Waterman	38	6	I, L	100	NA	NA	NA	NA	Drove through 6 V-belts
Akron Funk	39	4	I, L	63	2125	3.9 (98) × 4.3 (108)	201 (3.1)	260 (118)	
Rolls-Royce Vulture	39	24	X, L	1845	3000	5 (127) × 5.5 (140)	2592	NA	
Allison V-3420	40	24	W, L	2600	3000	5.5 (140) × 6 (152)	3420 (56)	2655 (1207)	Maximum hp was 3000
Lycoming XR 7755	41	36	R, L	5000	NA	6.4 (162) × 6.8 (171)	7755 (1264)	6050 (6050)	Developed from earlier XH 2470, 2400 hp
Chrysler IV 2220	41	16	V, L	1800	NA	NA	2220	NA	
Pratt and Whitney X-1800	38	4	H	2000	NA	5.3 (108) × 5 (127)	2599 (42.6)	2400 (1091)	Military designation was H-2600
Pratt and Whitney H-3130	38	24	H, L	2650	NA	6 (152) × 5.5 (140)	3730 (61.1)	3350 (1523)	Note odd discrepancy between H designation and displacement
Studebaker H-9350	4x	24	H, L	5000	NA	8 (203) × 7.8 (197)	9350 (153)	NA	Biggest ever; probably never completed; cylinder bore too big for flame travel and good ignition
Rolls-Royce Exe	38	24	X, A	1200	NA	NA	1342 (22)	NA	
Rolls-Royce Eagle II	44	24	H, L	3500	3500	5.4 (137) × 5.1 (130)	2808 (46)	3900 (1771)	

B—barrel; Ro—rotary; A—air-cooled; L—liquid-cooled; R— radial; I—inline; Iv—inverted; TS—two-stroke; O—horizontal opposed; V—V-type; X—X-type; W—W-type; H—H-type; NA—not available. Open figures—U.S. Customary; figures in parentheses—metric. Weights are dry.

Index

NOTE: Page numbers in **boldface** indicate illustrations.